Tony James is a retired police officer who lives with his wife and son in the northeast of England. He enjoys travelling, writing, playing musical instruments and reading. This publication is his debut non-fiction book and is based upon actual events that occurred during travels with his wife following their marriage in 1971.

This book is dedicated to my wife, Iris, without whom none of it would have been possible.

Tony James

Do You Want the Shortcut or the Scenic Route?

A History of Amusing Travels with a Wife

Austin Macauley Publishers™
London • Cambridge • New York • Sharjah

All of the events in this memoir are true to the best of the author's memory. The views expressed in this memoir are solely those of the author.

A CIP catalogue record for this title is available from the British Library.

ISBN 9781035847549 (Paperback)
ISBN 9781035847556 (ePub e-book)

www.austinmacauley.co.uk

First Published 2024
Austin Macauley Publishers Ltd®
1 Canada Square
Canary Wharf
London
E14 5AA

I would like to acknowledge the assistance given to me by the staff of Austin Macauley Publishers. Without their continued support and motivation throughout the publishing process, this book may never have reached a conclusion.

Table of Contents

Preface

There comes a time in everyone's life, even for a home bird like me, when the urge to travel comes calling. To alleviate the symptoms, a holiday is usually prescribed and the more adventurous the destination, the greater are the remedial effects that transcend upon the individual. So sayeth the Greek Philosopher 'Holiditus'. This is actually a load of b****cks, but I am sure you can catch my drift.

At this point, I should mention that the vacations recalled in this book are but a few of the many that have been taken since my marriage to Iris, who has been my partner now for what seems like forever. Iris, as her name suggests, is from a long line of family botanicals as she had a nanna called Ivy and great aunts named Rose, Violet and Daisy.

When she was born (in hospital), her mother had a vase of iris flowers beside her bed and so the name, which in Greek means rainbow, was chosen. She often relates this to friends with the comment, "Thank goodness they weren't chrysanthemums."

Anyway, I digress. The tales of foreign soirees that follow are a distant cry from those taken whilst single. Singles holidays have their advantages but these are mostly confined to the tedium of boozing, night-clubbing, gambling, more boozing and talking about football or girls in any order. Boy, am I glad that's all behind me now. ("Liar, liar your bum's on fire.")

Thus, the tales related in this book are some of the more amusing happenings, that I can still recall, and most of the people spoken of are identified by their real names. Those that are not have had theirs changed or concealed for fear of lawsuits. However, they do nevertheless exist (or are now dead) and all events actually happened, albeit that the order in which they occurred have been left to artistic licence.

Those that I have been able to trace have been consulted and all have given their consent as to the publishing of those events that concern them personally. In the case of the persons from the spirit world, I did a **'rare'** thing and consulted a **'medium'** in order to receive the affirmative two knocks. So **'well done'** there then. You have to when there is so much at 'steak', sorry, stake.

I am naturally indebted to them all and both Iris and I feel privileged to have met each and every one of them for they have brightened up our lives in so many ways.

Fortunately for me, I kept a log of some of our travels, at the various times they occurred, and it is only as a result of these scribblings that I am able to recall (and relate) the more amusing events that befell us during those halcyon days.

I earnestly hope that after reading these various accounts, your own lives may too be enlightened (and humoured) by what you will soon know. As they often say, "Give it a go, for you never know."

Chapter 1
Et Balearics a Tois, Mon Ami

It is only fitting that I start this journal, of our travels, with a story based upon a holiday taken in 1975 to the Balearic Island of Minorca in the Mediterranean. This was our first trip abroad (since we got hitched four years earlier) and at the time, I wondered whether this would prove beneficial to our relationship or bring about a quick divorce.

Could we endure each other's company for a whole ten days or would the strain become too great and erupt into violent domestic arguments? Only time would tell.

Being married and only seeing each other after work each day was one thing but 24-hour coverage for 10 days was something as yet untested. This was going to be a true test of our stamina.

*****

For various reasons, the flight we took was from Luton Airport at the beginning of October and upon arrival in Minorca, we were well pleased that the weather was in the mid-eighties (Fahrenheit) and when we eventually reached our hotel, the sea was still in its summer plumage of bright turquoise set against a beach of fine white sand.

We were staying at a resort called ‘Cala Galdana’ and the hotel appeared to have been built upside down (against a tall cliff) with the entrance, reception and swimming pool at the top and all respective floors leading downwards towards the beach at the bottom of it. We were lucky in that we had a room halfway down and that our balcony ‘overlooked’ the sea and had a wonderful view of the horse shoe shaped bay that made up the resort. Alas, the hotel also overlooked other issues such as basic hygiene in the communal areas, good service and edible food. Thankfully though, there was a working lift.

Despite these issues, we quickly got into our stride and checked out the local surroundings as well as hiring the cheapest car we could find for the duration of the holiday as we intended to see as much of the island as possible. I use the word ‘car’ in its loosest sense for it was a really compact ‘Seat 500’ with a long hire history judging by its general condition. It seemed to be crying out for a service although, one of the burial-variety, rather than a mechanical one. However, it was to surprise us on a few occasions and I will relate some of these events at the appropriate juncture.

We did not venture far for the first day as I was a bit apprehensive about handling a ‘left hand drive’ on the wrong side of the road for the first time in my life.

I can recall that on the first day we had it, I drove around the hotel car park at least seven times testing the brakes, steering and suspension. I had to park in several different bays during our stay because the engine leaked oil and was leaving black stains that resembled mini crop circles.

When we did eventually venture out onto the open road, it was an exhilarating experience to go beyond 5 mph and in a straight line. In fact, we were able to reach a flat-out speed of around 25 mph as the roads were quite quiet at that time of the year.

As we explored the nearby area, we spotted a local restaurant named 'El Pollo' (which translated means 'The Chicken'). We stopped off to christen the car park with a crop circle of oil and sampled the menu. The establishment also had an external cactus garden accentuated with a maze-like pathway.

We learnt that the owner was Minorca born and that he had invested all his life savings into a business which he believed was destined to be the start of a chain of restaurants not too dissimilar to 'Kentucky Fried Pollo'. As we enjoyed our first meal there we decided to revisit the establishment and it soon became a regular haunt of ours. I even started calling the owner Senor Ken but he failed to grasp the significance of such a gesture and simply smiled at me thinking that I was paying him a compliment.

Our life at the hotel jogged along at a leisurely pace and most of our time was spent sunbathing and swimming in the sea, as it was considerably warmer than the pool which was always in the shade. The only thing guests used the pool for was for cooling their drinks and feeding the penguins. I may have lied about the penguins.

The hours drifted into days and the days into more days of semi-hedonism before a new party of holiday makers arrived to replace those that had left. Although many were of Germanic origins, we found ourselves partnered in the restaurant by an English couple from the Suffolk region. They were called Brian and June and they were celebrating their first holiday abroad as a married couple just like we were.

This made for an instant friendship and we were soon conversing as if we had known each other for years instead of just a few minutes in reality. Brian had announced that he was a professional jockey ('horse' not 'disc') and June obviously liked the way he rode (a vain attempt at a married joke).

As a result of our mutual friendship, we all started to hang out together around the hotel and in touring the island in 'El Coche' (The Car). Then on the second day of their stay, we suggested to them the idea of us all going out for an evening meal at 'El Pollo' and they agreed.

At the allotted time, the four of us set forth in the 'Seat special' and travelled at speed to the restaurant which was located about five kms from the hotel. Upon arrival, I parked in an oil stain free zone and we all swaggered into the restaurant and found ourselves a suitable table before ordering a large communal jug of 'Sangria'. After downing that and buying another, we decided to order a meal of 'barbecued rabbit' or should I say 'conejo asado' (impressed, eh!) as the ones roasting on the open grill looked very tempting.

As the night progressed, we decided to go the whole hog and order some Spanish 'champagne' which, in reality, was just fizzy wine but was nevertheless hitting the spot as both the girls were starting to giggle and the conversation was getting louder by the glassful.

At this point, a posse of four well-dressed gents entered the dining area and occupied a table not too far from ours. It appeared that they were discussing some business deal and after eating their meals, they carried on drinking what looked like champagne. Our raucous laughter soon caught their attention and they invited us over to their table to share their bottles of plonk.

We quickly learnt (owing to their broken English) that two of them were from Paris (France) and were in Minorca to buy shoes from, presumably from the other two who owned shoe factories on the island. Minorca, in case you didn't know is famous for shoe manufacture as well as gin production. The two locals were obviously treating the Parisians to a night out in the hope of gaining a substantial order for their respective factories. It turned out that the drink we were offered was vintage champagne and I swear that I saw the two Minorcans wince as we started quaffing it back at an alarming rate. They must have been desperate for the business as they did not say anything. Besides, who were we to turn down their hospitality and they all seemed really nice people despite the fact that they did not really understand my north-east accent.

They thought that the word 'div'vent' (meaning don't) meant 'divine' (from the French word 'divin') or even 'amuse' from its Gallic equivalent 'divertir'. Even Brian and June were struggling so it all made for a promising night. As you can imagine this led to some strange conversations although we did somehow manage to communicate and also have a good laugh.

This, however, did not last long as the larger of the two Frenchmen was getting very drunk and somehow assumed that Mrs J was willing to allow him a kiss. I had to tell him she was married to me and that meant 'div'vent' touch. Well, I do not have to spell it out but his amorous intentions became more obvious and when she said that she had to go to the toilet, the swine got up and followed her.

I assumed, because I too was feeling the effects of the alcohol, that he was going to the toilet for ‘Hombres’ but a scream from Iris brought me quickly to my senses and I rushed out into the cactus garden (that led to the toilets) to find him chasing her around the twisted and maze-like footpaths. She managed to eventually locate the ‘Mujeres’ (women’s) loo only for him to follow her inside. Another scream caused me to run in after them both and for me to grab him by the collar before he could plant a smackeroo on her lips. I duly marched him outside and told him to ‘p*ss off’ otherwise I would render him unable to procreate any more French babies. He assumed I meant ‘procureur’ (meaning prosecute) and he uttered something in French that I did not understand but took to mean something akin to swear words. I naturally retorted ‘et Balearics a tois, mon ami’ (or something similar) as he wandered off to join up with his friends as if nothing had happened.

Mrs J managed to use the loo unassailed and upon returning to the restaurant, we all parted company with the businessmen and found a seat (a real one-not a hire car) well away from them and carried on drinking. My bladder, by this time, was near capacity so I set off to find the gent’s loo. Needless to say, that due to my condition and the fact that it was really dark, I failed in my quest. It struck me that this must be some kind of joke by Senor Ken, so I decided to water his cactus instead. This tactic I may add is akin to having a barbecue in a nudist camp and you need your full wits about you. Although swaying heavily on my feet, like most drunken men do, I managed to complete my task without injury.

Eventually, the night drew to a close and June had to drive us back as she was the only one sober enough to accomplish it. She did make one or two remarks about the car's handling but we all told her to stop whingeing and get us back in one piece.

The next morning Iris and I had the most horrendous hangover which we blamed on the barbecued rabbit. Having missed breakfast, we had to settle for a few coffees at the bar to convince our bodies that we were still alive. The rest of the day was spent lounging around the beach, eating ice cream and dare I say it, a few drinks to stop our bodies going into shock mode. We did not see Brian or June again until dinner in the hotel restaurant where we regaled the events of the previous evening. After our meal, we all became acquainted with an elderly couple called Geoff and Irene and we managed to socialise for the remainder of the evening in the bar area. Before retiring, for the night, we all agreed to do something together the next day.

Chapter 2
Hitting the Heights

When morning eventually arrived, the weather was cloudy, and a lot cooler, but despite this we all agreed on a visit to the 'Sanctuary of the Virgin' (probably a convent or a monastery) that was built in the seventeenth century on the ruins of a former Gothic church at the top of Minorca's highest (and only) peak 'Mount Toro'. However, we had a problem as our hire car would only take four adults at a squeeze and not the 500 it foolishly stated on its name plate at the back. Brian came to the rescue with a brilliant idea that he and June would hire a motor scooter (sadly, they did not rent horses) and follow behind us which is what we agreed on.

Iris was to navigate for me whilst Geoff and Irene squeezed themselves into the rear seats and duly imprisoned themselves as the car had only two doors. The going was slow due to the increased weight of the passengers but we did manage to attain sufficient momentum to reach 20mph. Brian and June were having some trouble with the scooter but did manage to catch up and before long, we all reached base camp at the foot of the mountain where we planned our motorised ascent of the terrain with its twisty hairpin road that led up to the monastery.

I suggested that Brian and June go on ahead as it was likely that we would have some difficulty with our speed up the steep winding route and its treacherous bends. Oblivious to the inherent dangers, I revved the engine to make us sound keen and off we set to negotiate the first incline. This was traversed with little difficulty and as such it gave me renewed confidence and faith in the car for us to proceed further up the mountain road. Irene was not so confident and enquired if we could make it to the top. I did my best to reassure her by stating that as long as we have oil in the sump, we would get there. This did little to alleviate her fears as a furtive glance in the rear-view mirror confirmed that she had a vice-like grip on Geoff's arm severely restricting the blood flow to his vital extremities.

Unfortunately, at this time, we had to make a sudden lurch towards the mountain's edge to avoid a collision with a speeding car coming downhill. This only added to Geoff's discomfort as Irene moved her grip from his arm to one of a headlock that Mick McManus would have been proud of. As I made the appropriate course correction, the pair of them were thrown against the side window before yelling at me and pleading for me to stop and let them out. Although I was somewhat hurt by their comments, I had to pacify them by stating that I needed to find a suitable level area for this to be accomplished. Sadly, for them, the mountain road did not level out until we reached the final bend near the summit.

When we did reach that location, I spotted a place to pull up and came to a stop where I instinctively applied the handbrake. It was only then that I discovered the brake did not work and we started to roll backwards towards a sheer drop. To add to the problem, the engine stalled and the use of gears proved futile. My only option was to reverse steer into a grassy verge and pray that the car would somehow come to a stop.

As if by magic, my prayers were answered and we managed to stop the car. After making a quick glance in my rear-view mirror, I caught sight of Irene's face and, well call me old-fashioned, but I would hazard a guess that she was less than well. Either that or sheer terror had turned her a funny colour which was closing in on a greenish-white hue that hobby artists like myself would call 'Eau de nil'. Having released them both from their confined space, it was agreed that we all travel on foot for the remainder of the route and I am pleased to report that this short burst of pedestrian energy enabled Geoff to get his blood flowing freely again to his arm and neck. Irene's facial colouring took considerably longer.

The monastery was delightful and after a suitable photo shoot, of the stained-glass window, I spotted Irene seated inside and appearing to be in prayer. I do not think that she was a religious person more likely she was seeking divine intervention for our return trip. Afterwards, I had the unenviable task of persuading everyone to get back into the car for the return journey. Judging by their collective expressions you'd have thought that I was asking them to form a suicide pact. People can be so hurtful at times like these. Anyway, after much persuasion, we all got back into the car and I started up the engine. Then by sheer skill and deft handling, I conducted a fairly tight three-point turn enabling us to face downhill and negotiate the first of the many hair pin bends. Going downhill was a new experience for me in this particular car and I was extremely impressed with the speed we were able to attain. I do have to admit however that I was alone in this thought, as Irene had resumed her vice-like grip on Geoff's upper limb.

As we started hurtling down the mountain, my mind turned to Newton's second law of motion which states *that the amount of force, that makes an object accelerate, depends on the mass of the object and the strength of the force*. Now it does not take a degree in atomic physics to work out the outcome, that an overloaded 'Seat 500' with poor brakes, and travelling down a mountain road, is the gravitational equivalent to a space shuttle re-entering the earth's atmosphere without the benefit of rear thrusters or a brake parachute.

I am pleased to report that as I am writing this historic account, you will no doubt have gathered that we made it safely back to base camp (albeit quivering) where we were reunited with Brian and June. There was a small crowd that had gathered to watch the spectacle and they managed a round of applause after we had alighted the car. Boy, did that make me proud! They must have thought we were some kind of adrenaline junkies.

Geoff and Irene stated that they would prefer to walk back to the hotel but after some gentle persuasion, she agreed to head back with Brian on the scooter whilst June took her place in the hire car. The whole experience had totally unnerved her and neither Geoff nor her spoke to us for the remainder of the holiday and, in fact, did their best to avoid us around the hotel. I was not particularly troubled by this and consoled myself with the thought that it's their loss not mine. The following day was at leisure around the beach area where we plotted our next soiree which was to explore the north of the island.

After breakfast, we set off in the car and to our amazement found that the main roads were lined with soldiers armed with machine guns and other suitable weapons. Iris remarked that word must have got around about our previous motoring exploits and the large number of oil stains that were mysteriously appearing around the island. I accordingly drove very carefully and did my best not to look them in the eye. We continued on until we saw a sign for 'Fornells' which at the time was a small fishing village situated in the north-east corner of the island. After turning off, onto a country road, the military presence disappeared and the rest of the journey was a tad more relaxing.

When we eventually arrived, at our destination, after stopping for lunch in a delightful little café, we made the decision to visit the capital of the island 'Mahon'. After consulting our map, we duly set off down the east coast route and eventually arrived on the outskirts of the city where we parked up and made the rest of the journey on foot. Our intention was to explore the shops and other tourist highlights that included the famous quayside where Admiral Lord Nelson had once docked. This main port area also housed many gin factories.

One of these establishments was encouraging passers-by to engage in a tour of the factory to see how gin was made and to sample some of the produce at the end of the visit. This seemed an excellent way to pass the time in the heat of the afternoon and in we went, along with some other stragglers who were obviously alcoholics like ourselves. Once inside there was no mistaking where you were as the smell of juniper berries pervaded the air. The guide was multi-lingual and I have to admit that it was an excellent tour and well worth the entrance fee of nothing. The best bit was at the end when we entered the tap room where the sampling took place.

We were all invited to taste a small glassful of the clear liquid. Until that moment I had never thought that neat freshly made gin, of around 60% proof, could taste so vile and my facial expression showed testament to this. A swift nod of the head saw the contents of the glass downed in one gulp followed by an exasperating 'phoaw' from my now severely disabled vocal cords. The other tourists who were present did something similar and the guide just laughed before uttering 'perhaps a little something to water it down' (in several languages). From the collective nodding of heads, this was unanimously agreed and the guide produce a bottle of a thick yellow liquid which he added to the replenished gin glasses. This time I decided to smell the contents of the glass and noticed that it smelled sweet and more inviting to the palate.

As before I raised the glass to my lips and took a swig downing it in one go. This time it took a little longer for my brain to register the outcome but when it did it felt like I had consumed napalm which had exploded within my stomach. I subsequently let out a muffled cry akin to a 'phoawawa' accompanied by a full-body shudder and a violent head shake. Mrs J had a similar experience.

The guide simply laughed out loud after explaining that the yellow liquid was banana liquor which in itself was over 40% proof. I then realised that he was not a guide at all but a sadistic misfit who would have been better suited as a torturer in Franco's army had he been born in an earlier time. The irony of it all was that he expected us tourists to purchase large quantities of the stuff and I sure as hell was not going to give him the satisfaction. Whilst some of the other tourists, in the group, were apparently negotiating a purchase price, we were able to sneak out and run for cover up a nearby alleyway. The rest of our visit to Mahon passed without incident before we re-located our hire car and set off back to the hotel.

Once we had left the city, we were once again confronted by soldiers guarding the sides of the road that led all the way to the island's other main city 'Cuidadela'. This time we were halted, along the route, with other traffic and had to wait until a cavalcade had gone passed. We later learnt the troops were present for a visit by King Carlos of Spain and not there to keep an eye on us. Greatly relieved, we eventually arrived back at the hotel in time for dinner.

The only other event, worthy of mention, was that I got stung by a wasp on my inner thigh as I was wearing shorts. Had the little bugger been a bit more adventurous then, I would have been truly incapacitated and unable to drive. The poison that it had injected into me was clearly making me delirious and my leg was swelling up and hurting like hell. Nobody at the hotel was willing to suck the poison out so I had to resort to drinking extra quantities of white rum and the hotels equivalent of coke to offset the effects. As you can imagine this took some considerable time and left me a gibbering wreck with bloodshot eyes and a furry tongue.

When the holiday eventually came to close, we wished our friends farewell and departed, as we had arrived, with bulging suitcases and very tired expressions. The flight back to England was uneventful and we touched down at Luton Airport on a very cold early morning. Our car would not start as the engine block had frozen and we had to pay for airport staff to give us a towed start to crank the engine back to life. Thankfully, we managed to get it going and drove back home.

The amazing thing was that we had survived our first holiday together as a married couple and that augured well for future expeditions to foreign shores.

Chapter 3
If You Go Down to the Woods Today! (Part 1)

I had this sudden urge. (Yes! another one. Possibly a man thing.) One wet Sunday morning in January 1977 to throw in the towel and start a new life elsewhere in a foreign land. I talked this over with Iris and surprisingly, she wasn't averse to the idea as, in fact, she had been thinking on similar lines herself. Nevertheless, she had reserved her thoughts until a suitable moment arose to discuss it. The problem was where to go, as we needed to future-proof for a possible family, and we would also need to consider such issues as education, health and the like.

Several destinations were mooted, including Australia and New Zealand, although surprisingly it was Canada that caught the imagination. This was due in part to us having friends living there and me just having read a book about a British family who emigrated under similar circumstances to our own. The book painted an accurate picture and included all of the hardships that were faced before they settled. I therefore commenced the task of finding out as much as I could about the country and suitable job opportunities. I was a serving police officer at the time and debated the possibility of transferring my role. In furtherance of this aim, I accordingly wrote off to several major city police forces and awaited a response.

I had no desire to become a Mountie (as I don't suit red), but simply to be one of the ordinary provincial police officers that control the municipal towns and city areas. Being a Mountie is more akin to being a federal police officer with duties that overlap strict boundaries. Hunting Indians, killing bears, riding a horse, or singing 'Rose Marie' to some native chick across a lake were not particular strong points of mine and at the time there were no evening classes at my local college for training in any of these special skills.

Canada had that frontier feel about it and it was known to be a clean country with a progressive government and therefore offered plenty of opportunity to budding immigrants, and so the die was cast. Between us, we devised a plan whereby we would combine a holiday with job interviews should the latter become a reality. To my surprise, and our equal delight, I was offered several interviews by different provincial city police forces so we decided to do the trip upon an ad-hoc basis utilising 'back pack' camping as a means of economising over our three-week trip. The first week would be surrendered to job interviews and the remaining two weeks to sheer holidaying pleasure somewhere in the Rocky Mountains.

The formalities were set up and (in June of that year) the great day finally came for us to fly out to this vast country from our local airport at Newcastle. The courier airline was 'WARDAIR' which had only recently commenced a direct flight arrangement as part of a new venture. Newcastle, at that time, was not normally geared up for long-haul destinations because of its limited runway space, so it was something of a bonus for us. As it turned out it proved to be even more so as we were treated to 'in-flight meals' on 'China plates' and ate fillet steak with real metal Georgian style cutlery. This was airline luxury and a far cry from our previous flight to Minorca where plastic everything was the norm.

The flight was a dream and when we touched down in Calgary Airport, in the province of Alberta the local time was around 2 pm. As we were travelling on a non-package holiday basis, we were subjected to more rigorous passport and immigration controls but eventually cleared the airport and took a shuttle bus into Calgary city. This was where my first interview was due to take place at around 4 pm and as such I had to use a public convenience to change from my travel clothing into my best suit carefully folded away in my backpack. Despite these problems, I managed to give a good account of myself at the interview and as it turned out, I was successful with the offer of a job and that they would be getting in touch with me at a later date.

Sweating and totally stressed out, I composed myself before returning back to the same public loo for a quick change into my travelling clothes. If the local vice squad had the toilet under 'observation' then I reckon I would have some serious explaining to do. After that, we headed across to the main bus depot to wait for a Greyhound service coach that would take us inland to the province of Saskatchewan. I was expecting some strange headline in the next morning's papers stating, "Quick change pervert sighted in city toilet."

It was of great relief when the bus turned up as it was getting late and we needed to get out of that place because we were being harassed by North American Indians and Eskimo 'drop outs' who were peddling watches and drugs. To be fair, I did not see any drugs although we were asked if we wanted some. Perhaps they were hidden inside the watches. It goes without saying that big cities are exciting places during the day but at night they take on a completely different persona and can be quite scary to the vulnerable tourist. We must have seemed like easy prey as we were carrying backpacks, spoke English and had that lost look about us. Maybe I should have rushed into the nearby toilet and changed into my 'Superman' outfit in an effort to deter them. Mind you, they probably had kryptonite watches as a safeguard.

I should mention at this juncture that I had made Iris's backpack as light as possible because she was three months pregnant at the time. This was something that we had discovered after we had booked the trip but decided to go anyway as the opportunity may not present itself again for a very long time. Although camping is not usually recommended for pregnant women, her pioneering spirit came to the fore (in the interests of economy) and we proceeded as planned. Most of our equipment was newly purchased and a **two-man** tent had been selected during an earlier shopping trip. It has always baffled me why they only build tents for men but perhaps it has always been regarded as a gay pastime. You can even get a **six-man** tent! The mind boggles.

We had passed this camping shop and were amused by the window advert which read, 'This is the discount of our winter tent' 40% of RRP. Yeah, I know the quote's been used many times before but the offer was too good to miss. We thought that if it was good enough for King Richard III, then it would be ideal for the wild terrain of Canada. We also bought a cheap camping stove, some insect repellent, a compass and extra tent pegs as all were on sale at the time. All of our other equipment was either hired or borrowed from friends. The actual tent proved ideal for our purposes and we became fairly adept at both erecting it and packing it away in ever-decreasing time limits.

Anyway, let's get back to the story. Our first campsite, which was on the outskirts of Saskatoon, Saskatchewan was something of municipal affair and far from the rugged wilderness I had envisaged. It was a perfect start as it gave us plenty of opportunity to settle into our new regime. Saskatchewan was a fairly flat province and two stops were required at both Regina (the region's most prominent city) and 'Saskatoon', it's more provincial counterpart. This latter location did not have the feel of a big city and we were most impressed by the cleanliness, modernity and beauty of the place. As it turned out, this place had that special feel about it and we both instinctively knew it was destined to be our new home if all went well at the interview.

It duly transpired that I was successful at the interview and made my mind up that this one was the one. So much so that I cancelled my remaining interviews at Regina and also at Edmonton, which was in the north of Alberta. We decided, instead, to travel to a place called Jasper which involved two train journeys via Edmonton.

Jasper was in the mid-range of the Alberta end of the Rocky Mountains and was more famous as a winter skiing resort than a summer vacation spot despite the fact that it was really picturesque. Both train journeys allowed us to gain sufficient sleep to rejuvenate our spirits and this in turn gave us the necessary impetus to trek to our new campsite that we had arranged at a place called Whistlers Mountain. This journey of around three kms was taken on foot along the side of Highway 93. Passing motorists were astounded to see two humans strolling along, as everyone and their granny drove everywhere even if they went camping.

We didn't really care what they thought as the weather was glorious and walking in the Rocky Mountains was something we had been looking forward to ever since we left England. Travelling by car is without doubt more speedy and more comfortable but you miss so much of what nature has to offer. This was an experience neither of us wanted to miss.

Along the route, we passed over the 'River Miette' and spotted loads of unusual wildflowers that lined the grassy areas between the road and the forest. The river provided me with the chance to empty my bladder, and in order to facilitate this, I clambered down the bank adjoining the road bridge and disappear from the view of passing cars and trucks. As I was enjoying a much-needed 'pee', I caught sight of something large and moving out of the corner of my eye. It was a fully grown elk that was having a drink on my side of the river. The newly formed waterfall caused it to look and stare in my direction, which in turn caused me to pee much harder than usual in an effort to finish the job quicker and get the hell out of there. Luckily, it was positioned upstream of me and my additions to the water level were being carried away in the opposite direction. It would have been extremely bad luck on my part if this had been the rutting season and the elk had been gay.

Having regained my composure, I swiftly put my tackle away and rushed back up onto the roadside where Iris was picking some unusual wildflowers. Unbeknown to her, a young elk calf had wandered out of the forest and made straight for the flowers that she was proudly displaying in her hand, and accordingly started munching them. In my efforts to save her, I took out my camera and took a lovely photo of them both. This must have unnerved the young elk as it turned and made off back towards the forest where it presumably re-joined its mother (or father) that had been having a drink earlier near me. The Canadian Elk (or to give it its other name 'Wapiti') is the largest of the deer family *Cervus canadensis,* with large much-branched antlers and is native to North America.

After all that kerfuffle, we set off to locate our campsite for the night on Whistler Mountain. Upon our arrival there, we were issued with our permit and handed a leaflet giving advice about encounters with bears. This was for both the 'black' variety (Euarctus Americanus) and its big brother the 'grizzly' (Ursus Horribilis). Did you know, for example, that it is futile to run away from bears, or climb trees, as they can out do you, on both accounts? Apparently (so sayeth the leaflet), you should curl up into a ball and pretend to be dead as that way the bear will lose interest after it has had a small bite out of you (assuming you don't cry out in agony). Our only advantage is that we have better eyesight than them and therefore it was a great comfort to us to know that we could see more clearly whilst being eaten alive.

Being the pioneer that I am, I dismissed these fears with the thoughts that, as this was a commercial campsite (protected by Mounties), we were unlikely to encounter any of these furry beasties as they would resist the temptation of wandering into a heavily populated camping zone. **Wrong!**

The campsite was pure wilderness save for raised wooden areas for tent pitching. The problem was that I did not pack hammer and nails to assist with erecting our type of tent and had to settle for a fairly flat piece of ground nearby that tent pegs could penetrate. The forested area was broken only by pick-nicking areas where rough barbecue devices, wooden tables, bench seats and tall timber food protection poles (with ropes) were assembled. This latter device I presumed was a Canadian idea to enable campers to haul their fresh food up in bags to a height that even the hungriest of bears would fail to smell the contents or bother to climb for it. This was a necessary rigmarole, if nothing more, than to deter bears from the main campsite area.

Funnily enough, it was not the bears that got to me, but the blasted midges. At the commencement of twilight, their 'command and control' centre must have sent out the necessary orders to attack and thousands of the little beggars were dispatched to the relevant areas where they could cause the greatest annoyance. One such spot was our personal campsite area and in particular our humble tent. Our only weapon was a packet of mosquito coils which you had to set fire to at the outermost tip, blow out the flame, and let it smoulder as it slowly burned towards the centre. This did nothing more than repel us from the tent. Canadian midges seem to be wise to these feeble British devices, either that or their air reconnaissance fed back the relevant intelligence and averted the swarm to a more promising target namely me. Much flaying of arms and other bodily aerobics failed to deter them and I began to pray for a bear as I thought the sight of one may have scared them off and allow me to get on with the cooking of our dinner.

Much to my disbelief, I heard a scream from across the clearing followed by the sight of two female campers running towards Mrs J and shouting incoherently in what sounded like a Yorkshire accent. As they approached us, I realised that they were indicating that a black bear had crept up behind their tent and was about to devour the food they had set out on their table for the evening meal. I did what any man would do, under the circumstances, and ran into our tent to get my camera. I approached as near as I dare (about 100 metres) and let off a shot with the flash on. The bear looked up at me but then carried on consuming their meal before disappearing as quickly as it had arrived. Iris was trying to console both the girls (who were in fact from Yorkshire) and upon my return, I announced that I had dealt with the situation and scared it off. I think they were mightily impressed by my sheer heroism and to show even more favour we offered to share our rations with them which they very gratefully accepted. After a few drinks, to steady the nerves, we set about hauling our remaining food up the poles before settling down to our first night on Whistlers Mountain.

As darkness fell, Iris was soon asleep and as tiredness crept over me, I was suddenly startled by a series of loud-pitched howls coming from the upper reaches of the mountain. These noises, which I initially thought were coming from the Yorkshire girls, caused me to instinctively grab my camera. The noises then turned into shrill whistles before becoming a total cacophony as other creatures of the night joined in and amplified the racket. Instead, I grabbed my torch and consulted our guide book and leaflet to see if I could identify the cause. It transpired that the sounds were coming from creatures called 'Hoary Marmots' and their strange calls was how the mountain and surrounding area had acquired its name. Sleep was surely going to be a problem for me but eventually, I dozed off.

Chapter 4
If You Go Down to the Woods Today! (Part 2)

The next morning I awoke with a sudden panic as there was daylight outside the tent and I checked my watch for the time. It turned out that it was only 6.05 am but, as my heart was pounding in my chest, I was now wide awake and ready to check outside the tent to see if anything had been disturbed or there were any signs of visitation from bears, hoary marmots, aliens or other forest critters. When I did eventually leave the confines of our canvas shelter, it felt as though I had entered a strange world where mist hung amongst the fir trees and a faint smell of woodsmoke filled the otherwise fresh mountain air. It was fresh, calm and above all eerily silent save for Mrs J's rhythmic snoring.

Suddenly I heard the sound of a twig snapping and looked to my right where the sound had come from. There before my eyes, about forty feet away, was a mangy old black bear sniffing at the air and peering at me on a frequent basis. At this moment, I recalled the warnings that my mother sang to me in the song, 'if you go down to the woods today' but sure as hell this was no picnic. As for the bear, I was not able to say whether it was a male or a female and I had no intentions, to go nearer, to determine its sex or for that matter to see if it had a button in its ear. I actually froze to the spot and tried to recall all that I had read in the 'encounters with bears' leaflet only to realise that my mind, by then, had become a blank. This was not surprising because at that precise moment, I could not even remember my name or why I was here in the first place. Then, as if by some shred of human self-preservation, instinct kicked in and the advice about rolling up and playing dead came to mind. The only problem was that if I curled up into a ball, the bear would know I was acting and would probably attack me anyway after awarding me marks out of ten.

As my life past rapidly flashed before my eyes, I pretended to be dead standing up. After all, Iris had told me on countless occasions that I was good at it every time there was decorating to be done at home. This then was to be my strategy and in a blind panic, I stood motionless whilst the bear edged towards me still sniffing the air in front of, and above, its own head. To my amazement, and utter relief, it suddenly veered off towards the food bags hanging up on the poles in the centre of the campsite and gave me the chance to edge backwards very slowly towards our tent.

I managed to safely enter the sanctuary of the tent and woke Iris to tell her that I had fought off a black bear and saved her from being attacked. She initially did not believe me until I told her to pop her head outside and see for herself. By this time, the bear had almost disappeared but regardless she did not believe my story despite the fact that I was still shaking and clearly traumatised.

To this day, I cannot recall being as scared as I was at that precise moment. I decided there and then that this was no place for a pregnant woman and I accordingly persuaded her to up-sticks and move on to more suitable accommodation despite the cost. Surprisingly, she agreed and after we de-camped, we made our way back towards Jasper in order to catch a bus to take us to the town of Banff Springs.

We were lucky that we did not have to wait too long and were soon ensconced aboard a Greyhound coach travelling south along Highway 93 through what is more commonly known as 'The Icefields Parkway'. Our bus driver turned out to be a very likeable chap who loved his job and gave all the passengers the benefit of his considerable knowledge of the route. He was, as it turned out, one of the best couriers we have ever known.

The trip was excitement all the way and we stopped regularly at scenic viewpoints that included waterfalls, rivers, mountain passes and wherever any unusual creatures became visible along the roadside. Of the latter, we were introduced to golden-mantled ground squirrels, mountain goats and even big-horn sheep. The parkway stretches for about 230 kms and passes through some of the most beautiful scenery in the whole of Canada. Those amongst you who have made this journey will testify to this statement. We even made a scheduled stop at The Columbia Icefield Centre where all the passengers had the opportunity to purchase tickets and take a ride up the nearby Athabasca Glacier in the boneshaking delights of a snow-mobile.

Other wonderful sights included Bow Pass, Lake Louise, Kicking Horse Pass and our favourite of all 'Peyto Lake'. You really have to see this with your own eyes to appreciate the phenomena that it is. From the parking lot at Bow Summit, a short trail leads you to one of the most breathtaking views that you could ever imagine. Far below the view-point is Peyto Lake, a seemingly impossible green/blue coloured expanse of water whose hues change according to the season. Before heavy melting of the nearby glaciers, in June to early July, it is dark blue. As summer progresses the melt water flows across a delta and into the lake laden with fine particles of ground-rock debris known as 'rock flour'. This remains suspended in the surface water and is responsible for the lake's unique turquoise hue as it reflects colours of the light spectrum with sunlight. The location was named after an early pioneer called Bill Peyto, who, accidentally, discovered it during an expedition in 1898.

I hope you are suitably impressed by that brief insight into one of Canada's main beauty spots because quite simply I have amazed myself. Perhaps I should become a travel writer and not waste my time with this useless drivel about our holiday experiences. Nevertheless, as I have started it, I should attempt to finish the story, so here goes.

Well, as I have mentioned earlier, our driver fancied himself as a part-time courier who enjoyed his lifestyle and making his journeys enjoyable for those who travelled on his single-decked bus. Towards the end of the journey, he started asking for the passengers to identify their nationalities and played a little alphabetical game starting with 'A' for Australians, 'B' for Belgians, 'C' for Canadians and so on. I whispered to Mrs J that for the purposes of the game we were white Zulus from Zanzibar in the hope that we would have reached our destination in Banff Springs before he actually got through the alphabet. No such luck as we featured under

'E' for English (rather than 'U' for United Kingdom) and decided to own up sheepishly as we were the only ones.

Once he realised that he had people on board from England, he immediately started asking some searching questions about life back in good old Blighty. One of his fervent and lasting pleasures (from his once-only trip there) was fish and chips and enquired if you could still get them out of 'The News of the World'. We acknowledged in the affirmative provided you visited the up-market establishments. He then bestowed the values of such an experience on the rest of the passengers who appeared deep in thought as to what the 'News of the World' was.

Before we arrived at 'Banff Springs', we had one final short stop at Lake Louise which is another gorgeous photo opportunity. Eventually, we arrived at our destination and after giving our driver a tip ('Kontiki' in the 3.40 at Doncaster), we set off to find the tourist information centre. The town of Banff was a strange place as none of its buildings were more than one-storey high with the exception of the Hudson Bay Trading Company that have the only and exclusive rights to build up to two stories in height. As a result, many of the businesses had basement sales areas.

We managed to locate the information centre and were able to obtain details of a suitable bed and breakfast establishment just north of the main shopping area. Upon arrival there, we were met by the owner, who was a woman of formidable stature. I say this in the widest of terms (if you will excuse the pun) because she was probably the fattest woman I have ever met in my life. Her name was Heather but I suspect that was wishful thinking on her mother's part at the time of her birth. To me 'Jabber the Hutt' (from Star Wars) would have been more fitting.

Despite her appearance, she turned out to be a really nice person who made us feel welcome and offered us a cup of coffee and some home-made cake. We felt we had clicked here and glad that it was to be our base for the rest of the holiday. Afterwards, we were shown to our room where we each had the luxury of a steaming hot shower before settling down for a nap on a wonderfully comfortable bed. When we awoke, it was early morning and neither of us could believe we had just slept for 12 hours non-stop. Iris blamed me and I blamed the bear and the hoary marmots as it seemed only right.

After breakfast, we set off to explore the town as well as taking a cable car trip up 'Sulphur Mountain' which is a popular tourist spot despite it ponging of rotten eggs at the summit. Another of the attractions was a boat ride on 'Lake Minniwanka', meaning 'lake of the water spirit' (and not the name of a small Indian brave who was fond of self-abuse). The lake was really a reservoir built around 1912 and forms the largest body of water in the Banff National Park. Our boat guide informed us that it was impossible to drown in the lake as you would freeze to death first. Well, at least that was reassuring. Other trips included scenic walks in the various parks as well as along the Bow River. Here we passed its small yet nevertheless spectacular falls which lie just below the famous Banff Springs Hotel.

I have to say that we thoroughly enjoyed our time spent in Banff but all good things must come to an end, and we soon had to take yet another bus back to Calgary for our flight back home. Upon reflection, we both agreed that it was a truly fantastic experience with the future prospect of us emigrating back before too long. Sadly, a few weeks later, I received a phone call from The Saskatoon Police to inform me that the Canadian Immigration Service had recently changed their policy on immigrants occupying certain jobs (and in particular the police service) unless the vacancy could not be filled by a Canadian national.

I had therefore to resign myself to the fact that our expedition had been a truly wonderful holiday that would be very difficult to top. Nevertheless, we decided to try and set to looking for pastures new and subsequently more adventures to talk about.

Chapter 5
Hey, Mon. You Want De Shortcut or De Scenic Route?

Due to the birth of our daughter Caroline (in 1977) and our son Andrew (in 1984), we were unable to take any more foreign holidays for quite some time. We did however manage quite a few 'UK' breaks but these were fairly uneventful and would not add any useful dialogue to this book. Well, that is until much later when a boating holiday on the Norfolk Broads and other foreign soirees were attempted. More of that later.

Fast-forward to 1991 when we felt that the time was right to leave the kids at home with Iris's mam and give ourselves a chance of another break on our own again. On this occasion for some unexplained reason, we felt that a Nile cruise in Egypt would be a very unusual vacation and as such the die was cast.

A Nile cruise is one of those holidays that not only sounds exotic but is (unless of course if you are an Agatha Christie fan) and I was more than glad that we had finally decided to pluck up the courage to go, despite the warnings of incredibly high temperatures, numerous biting insects and maybe, just maybe, plagues of locusts. It was intended to be a special holiday to mark a special occasion, namely our twentieth wedding anniversary, and although everything looked set for this ancient millennia location, we were struck by a bombshell that caused us to re-think our plans. Very rarely do we have to reconsider our ventures but alas, this was forced upon us by the first gulf war that was escalating at the time and threatening to involve all Arab nations into the fray and that naturally included Egypt.

And so it was back to the travel agents and that laborious task of wading through the offers as they were presented upon their computer screen. As we were at a bit of a loss to find somewhere to replace or match our dreams of an exotic holiday, we had become a source of irritation to the lady travel agent who, as it was late in the day, was obviously keen to get home. Two budding tourists with the world at their disposal, you'd have thought, would have provided an exciting challenge to the agent but seemingly it doesn't quite work that way. They like customers who have a fixed destination, fixed date, fixed price, fixed number of persons in the party, fixed board arrangements and fixed everything. Not to put too finer a point on it, we were the exact opposite.

Suddenly, as if to answer the travel agent's prayers, a new destination appeared on her screen, namely the Dominican Republic. This Caribbean destination had only just become available to the general British public and was being promoted for the first time that year with the incentive of an 'all-inclusive' arrangement. I have never been keen on 'Republics', and as this country shares an island with the voodoo nation of Haiti, I was a little dubious to say the least as was Mrs J. Prior to the amalgamation of these two countries, the island was collectively known as 'Hispaniola' which added to its mystique with visions of pirates and the like. As such our interest was sated and we duly paid our deposits and were looking forward to yet another grand adventure.

The very next day, I hurried off to my local library and after much searching eventually found a book with some limited information about our new destination. Added to this there was also a television programme, that featured the island, so we had an inkling of what to expect when we got there. From all the data we had gathered, one seemingly minor point stuck in my mind. This facet was that travellers to the Dominican Republic should be aware that if asked whether you wished to have your baggage carried or transported by ad-hoc locals outside the airport, you should refuse or you may never see your cases again. I, being a cautious person at the best of times and a serving police officer to boot, was in no mood to allow a simple theft to embarrass me in such a way. I therefore devised a strategy whereby I personally would take control of our baggage and guard it with my life until it was safely ensconced in our hotel room.

Besides this minor consideration, there was much preparation needed before this issue became of relevance. Amongst this was the fact that we would both be required to undertake a great deal of vaccinations, with this destination being regarded as a 'Third World Country'. Little did I know how this would affect me personally. It turned out that we needed Typhoid, Cholera, Tetanus, Hepatitis A as well as a course of Malaria tablets. It seemed to us to be a daunting task to gain immunity from 80% of the world's known diseases but we had little choice in the matter.

We bravely attended our set appointments (with the nurse at our local surgery) and took each precision jab in a strict sequence. All seemed well (in my case) until she asked me to drop my trousers and present my buttocks for the finale. This was the 'Hepatitis A' jab and at the time was the only method of inoculation. I did as I was bid and lay face down upon the medical couch. A sharp pain was followed by strange feeling that the nurse was attempting to fill my right bum cheek with a dense liquid that felt like mercury or something equally heavy. When she had finished, I quickly gained composure. I stood up in a single movement and adjusted my trousers as I did so. It was at this point that my right leg went dead and I naturally thought, *good grief, she's overdosed me and filled the leg as well as my bum*. Anyone who has undergone a similar injection will know exactly the experience I speak of.

Following these vaccinations, I felt invincible. Now I could walk free without fear of mozzies, dirty tea towels, burger bars, council tips, wasps, rusty nails, hospitals and any other general danger that usually accompanies our daily lives. I was 'Dettol Man' capable of killing any known germ that came within my intimate zone. Now that my bottom had finally got better and I was ready for anything.

When the great day arrived, we flew off from Manchester Airport without a hitch. The uneventful flight lasted a little over ten and a half hours with a re-fuelling stop at Bangor in the state of 'Maine', USA. For the uninitiated, the state of Maine is world famous for its lobsters although we never saw any during our brief visit. This is just as well I thought because, believe it or not, we were not vaccinated against them.

When we eventually touched down at Puerta Plata Airport, we got something of a culture shock. What I thought of as a Third World Country was more like a fourth. There was no shuttle bus to transport us to the terminal building (if you could call corrugated tin sheds a terminal) and all luggage was hand-loaded off the plane directly onto the runway tarmac. All passengers had to collect their respective suitcases and haul them to the 'terminal'. Wheeled suitcases were a luxury in those days and ours had none. So, without further ado, I grasped both cases solidly and with white knuckles gleaming in the afternoon sun I set forth, at speed, towards the forbidding looking building. Without realising, I was first to reach the entrance doorway with Mrs J in hot pursuit trying to keep up with me. She was shouting something at me but my mind was fixed and I failed to hear her plaintive cries of 'slow down, you effing idiot' or words to that effect.

The inside of the building was really dark and a strange smell hung in the air. All I could make out were the silhouettes of some people but little else as my eyes had not, at this point, fully adjusted to the darkness from the bright sunshine outside. Then this shadowy figure, with bright shiny teeth (they were the only things in focus) approached me and asked for my bags. I immediately recalled the advice I had read about and said, "No thank you, I can manage them myself," before continuing my charge towards two large doors guarded by men in military-style uniforms who were also carrying rifles. Seeing my white knuckles and the look of sheer determination on my face, they moved out of the way and allowed me to open one of the doors with my foot. To my surprise, this was the exit from the so-called terminal building and a mass of locals were waiting in the sunshine, obviously for porterage reasons. Many voices rang out in a form of broken English, 'carry yo bags, sir,' 'carry yo bags, sir.' I pretended they were not there and made my way towards an English-looking lady, dressed in what resembled a tour representative's outfit, and identified myself to her. It was at this point that I learnt that the man with the shiny white teeth inside was in fact a combined customs and immigration officer. **Whoops!**

As British tourists were still a fairly new entity, on the island, he forgave my small indiscretion and duly stamped our passports and allowed us and the other passengers to proceed, once again, into the bright sunshine. The laws of obscenity prevent me from printing what my wife said to me afterwards.

All of the passengers were then separated into smaller groups and directed to small minibuses for our onward journey to our respective hotels. We were in a party of ten people bound for the resort of Playa Dorado on the north coast of the island. Our driver identified himself as Peter and did his best to whip us, and our fellow passengers, into an early afternoon Caribbean frame of mind. This is not that easy when your brain is telling you that your still on British time, which by my reckoning was definitely late evening and way past my bedtime. He managed somehow to get us all singing some calypso tune before asking everyone, "Hey, Mon. You want de shortcut or de scenic route?" As everyone was clearly tired, and well pissed off by this time, we all foolishly, as it turned out, agreed upon the former of the two options.

Peter immediately turned right directly onto a rough track through a sugarcane field and like a maniac drove his minibus amongst the tall stalks without regard to what damage he may be causing to the crop or the land itself. We, and our fellow passengers, were thrown around as if on a fairground ride and we all feared for our live as no seat belts were provided and no handles existed to hold onto for added security. This mini bus transport had probably been fashioned out of driftwood (and other scrap products) and seemed to be devoid of any recognisable suspension, Peter thought this was great fun and his laughter drowned out the screams from some of the younger female passengers.

Eventually, the sugarcane track subsided and a small road emerged that was almost flat. This offered us all some brief respite before Peter plunged the vehicle back into another section of the crop field and more rugged terrain. We anticipated that if we had to endure much more, our holiday may have to be spent in traction in a local hospital or (at least) a few visits to a chiropractor assuming that they had either on the island. I think all aboard were secretly praying at this point for a normal road and as if by magic, we swerved out of the sugar cane field and there it was. Peter informed us that the fun was over and that we should be arriving at our resort in about another twenty minutes time. In pursuit of some relative calm, we pleaded with him to stick to the scenic route from now on.

Chapter 6
Dirty Dancing with a Mai-Tai

Once we had arrived at our chosen destination, namely 'The Corfresi Beach Hotel', we were all greeted by a reception committee who met us all with a Hawaiian-type greeting by handing out 'lei's' (flower necklaces) and a rum cocktail for each and every one us as this was an adult only hotel. They then sang us some local welcoming song which after the bus transfer provided a heavenly interlude. Once concluded, we were directed to the main reception desk where all the men received fluorescent green wristbands and all the ladies received fluorescent pink ones. These we had clipped to our chosen wrist and informed that they were our resort passports to pleasure and must not be removed until the end of the holiday.

It felt like being in a hospital without restrictions and we came to learn that this was common practice for 'all-inclusive' resorts. Staff had strict instructions not to serve food or drink to anyone without a wristband and the armed guards at the entrances to the complex had similar instructions to shoot on sight. It all seemed a bit unnerving but was necessary for our own safety and security. Most of the local people living nearby were extremely poor and as such were prepared to sneak onto the resort and commit various crimes mostly theft. This could happen at any time of the day or night.

After settling in our room, we set off discovering the complex. The main building was like a giant conical building with a thatched roof made out of coconut tree fronds and between that and the beach there were three pools. The main one was filled with fresh water, and another was filled with sea water and had more buoyancy whilst the third was more akin to a lagoon that was fed 'tidal wise' from a rocky channel linked directly to the sea. The beach was really nice and the whole place had a Caribbean vibe which was something that was new to us.

Spanish seemed to be the main language and as I 'Hablo un poco Espanol', I was able to make myself understood to a lesser extent. I rapidly acquired the necessary lingo that was sufficient to order a whole lot of alcoholic drinks including a vicious cocktail known as a 'Mai-Tai'. One thing you learn quickly in the Dominican Republic is that the local rum is over proof and averaged around 150% which equates to 70% pure alcohol and needs to be given the utmost respect. The bar staff referred to this by the term 'gasoline'. If you wanted a white rum and coke you asked for 'gasoline and cola' which was probably equivalent to around four standard Bacardi and Cokes back in England. The Mai-Tai however was a different kettle of fish altogether and is not to be taken lightly.

I managed to acquire a basic recipe from one of the bar staff who told me it was made up of sizeable quantities of gasoline, with generous dashes of local over proof brandy, over proof triple-sec, over proof pineapple juice and a secret ingredient that he refused to divulge. This no doubt was over proof as well. It was a good job that the IRA don't holiday on the island because it would prove to be a much cheaper alternative to Semtex if they managed to find out what the secret ingredient was.

One apparently hardened drinker from Sheffield, who arrived on a later flight than ours (probably Gatwick) decided that he would show off his skills and decided to see how many Mai-Tai's he could drink. I'm not sure what the hotel record is but he was determined to break it. He never actually managed as he had to be taken to the local hospital with alcoholic poisoning, or at least that was the word that was going around after he was stretchered out of the bar area.

We did our usual 'let's get to know the other guest's routine' and very shortly assembled quite a number of new friends who turned out to be a good socialising group. This helped the holiday pass much more pleasingly. One chap I became acquainted with was a retired police officer from Barnsley called Bert. He was holidaying with his third wife, the other two obviously passing away through boredom or from being averse to his high consumption of beer. Bert was your typical PC Plod and had the characteristic round face, red nose and a pot belly brought on by years of after-hours drinking and socialising.

Now there is a poem by Rudyard Kipling called **'IF',** that starts:

If you can keep your head when all about you are losing theirs and blaming it on you.

This I would say pretty much summed up the situation at this time. I was getting blamed (by Bert's wife) for leading him astray and getting him drunk. Actually, to be more precise, it was more the other way around as Bert kept going to the bar and ordering beer for both of us and subsequently regaling me with his 'swing that lamp' tales that used to be a feature of the police officer in days gone by. The things he got away with during his service (on the beat) would make your hair stand on end especially if you were drinking Mai-Tai's at the time. I'll swear that his nose got redder with each pint but there again it could have been the sun, so perhaps it was dangerous to judge him too quickly.

There was also one of those couples who seemed to be all prim and proper but with a drink or two inside them they turned into right party animals. So it was, with Gordon and May who, were both endemic to the Wiltshire area of England and who both held professional positions in their jobs back home. They were a childless couple but I did not delve into the reasons why and only assumed it was because of May's strict upbringing and convent schooling that she told us a little bit about.

Despite this element, they seemed a likeable duo and attached themselves to our group for the entire holiday. I believed that it was because we had managed to meet up with a good crowd of people. It just so happened that we were also the noisiest and seemed likely to be the most fun. As a result, they were clearly breaking out of their normal mould, and were feeling somewhat wild and looking forward to risking whatever experiences came their way.

One night, after a few rum punches, we persuaded May to be bold and try a 'Mai-Tai' cocktail. She was reluctant at first but after considerable peer pressure from the rest of the group, she finally succumbed and sampled the delights of one. To her great surprise, she enjoyed the taste so much she ordered another. This is going to be fun, I thought, as I watched her senses become affected and her demeanour change quite dramatically. She suddenly grabbed Gordon and whisked him off to the dance floor as the DJ was playing one of her favourite songs. The rest of us sat back and watched them in an almost sadistic fashion.

Gordon was fairly well inebriated on the local beer and had shaken off all of his inhibitions. For a short while, on that dance floor, he transformed himself into bad boy 'Johnny' (Patrick Swayze) whilst May had become 'Baby' (Jennifer Grey) his Dirty Dancing partner. The other people on the floor made room for them and formed an orderly circle that gave the pair centre stage amongst a baying crowd of onlookers. Our boozing party had to stand on our chairs to get a better view. Everyone loudly applauded them at the end of the record.

Flushed with success, and a mild degree of embarrassment, the new 'king and queen' of the disco returned to their seats and both had suddenly become elevated to 'stars' within our group.

May was now in full swing and clearly looking to down another 'Mai-Tai' but sense and sensibility returned within her demeanour and she went back to drinking rum punches. Gordon was clearly still in a state of shock having discovered a whole new side to his wife that had obviously lay dormant within her. Either that or she had secretly been having Latin dancing lessons. He was, judging by the grin on his face, very pleased with this new persona and was contemplating his strategy for a wild night in bed and, as a result, he carried on drinking with renewed vigour.

May was putting herself about a bit and when it was my turn to be conversed with, she whispered in my ear that she fancied me. I told her that it was the drink to blame and that she would either forget this evening or regret everything should her memory return, which was probably unlikely after the two 'Mai-Tai's' and the rum punches she had consumed. She nevertheless carried on telling me (in a whispered voice) about Gordon's lack of ability in arousing her sexually and her frustrations of failed attempts to convince him that couples made love in more than one position.

I listened and pretended to be interested but she kept repeating most of the tale, in a slurred voice, that was becoming a little boring. I tried to switch the conversation around by telling her a joke about a horse that I had once bet on. The horse was called 'MFI' but I had lost my wager as three furlongs from the finish line one of its back legs fell off and, to top it all, the jockey was missing. She obviously did not buy furniture from MFI as she had no idea what I was talking about. Her musings about Gordon had seemingly reminded me of the joke but I cannot fully recall why. She did however giggle a little, as some drunken women do, and started to ramble on about something else so the tactic was successful.

The end result of the evening was that we all declared we had had a great time as well as a good laugh. We then dispersed and retired to our respective grass huts for the night. I remember that evening as being particularly humid and as I lay in bed perspiring heavily, due mainly to ineffective air conditioning, my thoughts drifted to Gordon and May who were no doubt feeling the heat as well, and that he might be having problems making bodily contact with her (in bed) particularly if she had applied her baby oil which she also admitted to me during our earlier conversations. With a wry smile etched upon my face, I slipped off into a blissful sleep ready to face the next day's challenges.

I awoke early the next morning with a severe hangover and it took considerably longer than normal to get my bearings and remember where I was. A quick peep out of the curtains reminded me that we were in the Caribbean and were about to experience another glorious day, weatherwise. I was also feeling very hungry and was anxious to avail myself of breakfast. Regrettably, I was about an hour too early so I got dressed and went for a walk whilst Iris slept soundly. I decided to go down to the beach through the hotel's gardens and by the various pools in the complex. The morning, in my opinion, is the best part of the day and many of the guests miss out on 'all-inclusive' packages because of their attempts to drink as much free booze as possible and end up comatose and hung over in bed until at least 10 am or later. Some do not even see the light of day until after noon.

When I did catch sight of other human life, it was only to claim specific sun loungers by registering their claim with a marker in the form of a hotel bath towel. This apparent obsession, made popular by German tourists the world over, was now adopted by British holiday makers who needed the security of their own familiar space in the most desirable sun spots. The guests I saw had clearly set their alarm clocks for a specific time and were dashing out half-dressed in order to locate the towels before dashing back to their sleeping holes where they could resume their sleep in the knowledge that all was well and that their sun loungers were reserved for when they were needed.

Failure to observe this routine by 9 am each day meant that all but the most decrepit of loungers were taken as well as the best spots for sunbathing, if that is what floats your boat. After my morning constitutional and returning to my little grass shack, there was already a sea of towels that had appeared as if by some mystical force. As most of the towels were identical, it was difficult to see how anyone knew whose was whose. I found the whole process to be a little childish but 'hey hoe' everyone to their own.

Although the Dominican Republic was officially a Third World Country, the complex we were staying in was luxury compared to the conditions that the nearby local people lived in. One organised excursion we took to Puerta Plata town was testament to that and showed us some of the squalor and basic conditions where people lived along the route that we took there and back. There was no proper sanitation, running water or mains electricity and their homes were built from scraps of wood and corrugated iron sheets. Afterwards, I am sure that those (like us) that took the trip felt somewhat guilty staying in our resort, by comparison, and the levels of decadence we enjoyed within.

Our complex had its own private bay and was a perfect setting for a stroll along the golden sands. It was to us a paradise island and an idyllic place to watch the sunset as it changed the colour of the sky blood red tinted with hues of both orange and purple. To make things even better, I was developing the best tan I had ever had in my entire life. I put this down to the warm Caribbean winds that blow in from the sea and help keep the temperatures down to within reasonable limits. Normally, I go pink then red then deeper red before my skin peels and goes back to a very sore pink ready to recycle the whole process.

Here I was turning a golden tan and in my own mind, I was resembling some Hollywood sex god with a six-pack to match. Mrs J soon brought me back to earth by letting me know that my pot belly only counted as a one-pack due to the high levels of food and drink consumed up to that point. Undaunted by her comments, I continued with my dreams and the hedonistic life style before it all came to an end and we had to prepare for our return trip home.

We saw out the remainder of our holiday in true Caribbean style until the day came and we had to return to the airport. Once there, I hoped that all the officials had forgotten about my indiscretions and allowed me to board without any problems. This they indeed did and before long we were airborne and heading for Boston, USA (our scheduled re-fuelling stop) and then onto Manchester, England. Regrettably, this was followed by the drive home to the north-east in a totally jet-lagged condition.

Chapter 7
The 'Kiddy' Years

In case you think that we are bad parents that always leave their children and jet off to warmer climes without them, then I need to set the record straight before you send a report to social services. We love kids (but neither of us could eat a whole one) and our two were the apples of our eye. We made sure, as they were growing up, that they had plenty holidays with us in the UK and even took them to various European destinations during school holidays.

Sadly, they did not always want to come along and often preferred to stay with their gran who spoilt them rotten with treats and other favours. Neither of them liked hanging around airports for long hours and neither did they enjoy any flight over four hours which seemed to be the limits of their confinement in a packed aircraft before they became restless and started moaning. The best we could manage was a holiday to the Canary Islands on at least two separate occasions. Once to Tenerife and the second to Lanzarote.

The trouble with family holidays is that you rarely get the chance to let your hair down with other childless couples and accordingly these breaks do not always present the opportunity for social escapades that we so often enjoy when we go as just a couple. However, there is one holiday that broke the mould and this will be added in a later chapter at the appropriate juncture.

It therefore became a balancing act for us and (granny permitting) we managed somehow to arrange and escape on holidays without them to more far-flung destinations, as when the opportunity presented itself. If you haven't yet ditched this book, then you may wish to read on for further adventures.

One such occasion was a chance to visit the Atlantic Island of Madeira for a one-week holiday in the late 1990s. On this occasion, the kids were of an age where they could be left on their own with granny as back up.

Chapter 8
In the Shadow of Helga

The island of Madeira (to us) had long held a mysterious air about it and we were keen to pay it a visit. It is an autonomous region of Portugal but generally more pricey and less populated (or it was when we went). It seemed to be more suited for the wealthier tourists with a limited number of hotels that meant fewer beds for general occupancy. Property prices were higher than most European countries and seemingly the islands' government wanted it to stay that way. Our chance to visit came out of the blue and only as a result of a last-minute cancellation by other holidaymakers.

We had little option but to fly, once again, from Manchester as the only other option was from Gatwick which was too much of an outward and homeward trip by car. The time of year for this holiday was June, a month regarded by many as the best time to travel there even though it was blessed with a year-round sub-tropical climate. Our good fortune did not end there for we had managed to secure accommodation at a 'four-star' residence, on a half-board basis, located beside a lido area just west of the capital Funchal.

This turned out to be one of the most popular and sought-after locales. We were starting to wonder when we were going to come across a hitch. Let's be fair, there is usually a hitch as good things seem to happen to other folks rather than us. Generally speaking, we encounter some form of a problem and end up paying the price as a consequence. The old saying goes that pleasure is often followed by pain but no matter how hard we tried we could not come up with any. For once we had to accept that we had been really lucky, at least, for the greater part of our seven-day holiday.

With the benefit of hindsight, I reckon that Madeira is a place that you need to visit at least once in your life and that such a visit should last longer than one week as there is so much to see and explore. Having said that, two weeks would be a tad too long so go for 10 or 11 days if you can. In our case, we were determined to see as much of the island as time permitted and take in the sights and sounds of this Atlantic jewel.

During our stay, there was a classical music festival in Funchal which appealed to me personally as this was one of my favourite genres of music. Mrs J on the other hand was not a fan as she preferred more lively stuff that you could dance to. I prided myself in the fact that I was, and still am, a dilletante of most wind instruments. Although my chosen favourite is the saxophone (of which I owned three, at the time, soprano, alto and tenor) I also play flute, clarinet, a penny whistle, for what it is worth (excuse the pun) and a guitar. Mind you, I do not get much of a sound out of the latter by blowing it. My paternal grandfather had been a professional musician and I suppose it was a second generation thing as my father did not possess a musical bone in his body. All I knew was that it came easily to me and I really enjoy making music as it totally relaxes me.

Before I get onto the finer details of our visit to the island, I think you ought to be aware the Madeira has the second most difficult runway in the world for a trained pilot to negotiate. This, was at that time, due in part to its limited runway length but mainly attributable to the mountainous terrain which makes for a tricky descent with a turn halfway down. They say that Hong Kong was worse because of the encroaching skyscrapers and I can actually vouch for that, from the time when we made a trip there at a much later point in our lives. Maybe more of that in another chapter.

When we did arrive there by plane, our landing was even more tricky due to the fact that they were extending the runway farther out into the sea and, by necessity, had to close off part of the existing one making the job of landing even more treacherous. It was a relief to know that our plane had decent brakes and healthy reverse thrusters to boot. You can't beat a good set of reverse thrusters as my old gran used to say.

We started our holiday with a reconnoitre around the lido area before branching out with a slow and leisurely walk down to Funchal. En-route we passed through a park that was ornamented with modern art sculptures, mostly of nude men. Whilst I was not generally impressed, the rear buttocks of one well-endowed figure caught Mrs J's attention, as well as her wandering hand, and I decided that this demanded a photograph for 'posteria-ty'.

We sidled down to the capital taking in the other sights of this cosmopolitan city with its colourful flower stalls and its vibrant quayside. I was particularly taken by the high number of ugly local women whose faces were etched like 3-D road maps. This I put down to the vast amount of exposure to the near-constant sunshine and the thin mountain air. Either that or years of in-breeding with orangutans had left its mark. Mind you in the interests of not wishing to appear sexist, I have to add that the local men were not much better looking. In fact, their faces were probably deeper etched than their female counterparts which made them appear even more wrinkly.

'People watching' can be a fascinating hobby but not one that I would waste much time on, particularly on a short stay holiday. I only mention it because it occurred to me that the locals were a tad more interesting to look at than the countless number of banana trees and the bougainvillaea that seemed to be growing everywhere. One of the other points of local interest are the roads which have to be the most winding in the world. They have virtually no straight sections or level sections due to the islands' rugged terrain.

I had originally considered hiring a car and tour around but after seeing the status of the highways and byways system I wisely dropped the idea. In fact, I think that I can confidently quote that one of the world's smallest books has to be 'Caravanning in Madeira'. I am adding that fact anecdotally on the basis that there are no trailers of any kind to be seen. Even if there were they would surely be limited to the quayside. Anyone who has visited the island will know exactly what I mean.

Despite its ruggedness, Madeira has a beauty and tranquillity that is hard to beat and many fine walks abound in the mountainous regions of the interior. These trails are locally known as 'Levada's' which are a combination of irrigation channels, and pathways, that were carved into the mountain sides centuries ago by forced slave labour. These are now a unique feature of the island feeding fresh mountain water to the crops in the low-lying fertile areas and are even used as a source of hydro-electric power. However, their main popularity with visitors is the fact that they provide some of the most beautiful and breathtaking walks that are to be found anywhere in Europe.

These 'Levada' walks have many natural features such as waterfalls and sub-tropical fauna and flora (flowers not warm margarine) and are well worth an excursion should you make a visit to the island. The mountainous interior is spectacular, especially a place called 'The Nun's Valley' although other fabulous sights abound. Alas, travel facts are not the main intention of the book but I thought that I would share that information with you for good measure.

Anyway, let's get down to more serious stuff and an insight into another established feature of this Atlantic jewel, namely timeshare. Yes! My friends, you cannot escape it. Here it is called 'holiday ownership' but it amounts to the same thing and sales touts are out and about along the main thoroughfares, including our well-trodden path between the lido and the city centre. Our first encounter was on the second day of our holiday when we had the pleasure of exchanging dialogue with a young chap who told us his name was Peter and that he was from Birmingham (England). He had a strong Black Country accent which set him apart from the other touts and his opening pattern was to spare him a few moments of our time to save his wife and children from starvation and/or the dreaded social services. Well, I ask you, how could you not stop and talk to someone who was as upfront as that?

What gets me is how these touts know you are Brits before you have uttered a single word. Even when I pretended to be German with a reply of 'Der stranden von oben gesehan' (the only German 'sprecken' I know) which roughly translates as 'The beach as seen from the sea' (this was off a postcard I had purchased many years earlier in Spain). He cleverly saw right through this masquerade and replied, "Yah, yah, now as I was saying," and then rambled on about the virtues of this nearby complex which had units available for holiday ownership. The crux of his sales pattern was that if we just went along for the demonstration sales pitch, he would get paid, his wife and kids could get fed, and we would be able to claim a couple of free tickets for a jeep safari into the mountains. Simple, eh!

I have never been keen on timeshare, or holiday ownership, or whatever, although Mrs J liked the challenge of pitting her resilience against the hard-sell salesmen, and after leaving them verbally worn out, walking off with her promised reward. Despite my protests, she succumbed to Peter's social begging and persuaded me to tag along for the experience. We were then guided, by him, to a rather posh hotel complex where we were requested to take a seat. Peter then left us to return to his designated pestering spot waiting for more willing victims of his sweet patter.

After a short wait, we were informed that we would be met by someone who would take us through the concept of holiday ownership. My attempts at injecting pessimism into the proceedings were shot down in flames by my wife who told me to stop being a 'wimp' and to brace myself for some action. I was not convinced but decided to put on my meanest-looking face whilst several men and women passed by us but none stopped to move the proceedings onto the next phase.

Just when I was beginning to regret this total waste of our holiday time, a tall smartly dressed lady approached us and asked for our first names which we supplied to her. She spoke in English, with a German accent, and introduced herself as 'Helga'. At this point, I decided to remove my sunglasses (that I had kept on to maintain that super cool look) and was immediately taken aback by her attractiveness and very desirable figure. She had my undivided attention from that moment on.

Helga invited us to follow her to another part of the hotel and showed us a model of the new complex that was on offer for holiday ownership. As she was outlining the details of various units available, I had the chance to eye her up and down at closer range and quickly formed the impression that she was in the category of 'shag-a-del-ic' although I did not make this obvious to Mrs J who kept nudging me in the groin for some unknown reason. Helga was very tall for a woman, being my height at least, and I'm 6'2" with my socks on. This did not deter me as I was more transfixed with staring at her bosom and trying to decide if she was wearing a bra or not.

I came to the conclusion that she was not, judging by the way they moved under her loose-fitting top. A couple of points she was putting forward were slowly hypnotising me into signing up for at least two weeks of the scheme. That was of course until another swift jolt in the groin brought me back to my senses and caused me to replace my sunglasses to hide the tears that were forming in my eyes. She (my lady wife) had caught me good and proper on this occasion and I did not want Helga to be aware of this. Look cool and stay cool was my motto.

We did, however, agree to be taken across to the actual site and be shown around both the completed apartments and those still under construction. We had to wear white plastic hard hats (for safety reasons) and if I say so myself, I rather suited mine as it gave me that rugged 'Bob the builder look'. The tour lasted about twenty minutes before Helga took us both to an office complex where she showed us a plan of the remaining plots available for purchase. At this point, you have to decide, rather hurriedly, how you are going to make your escape, for it is here that the really hard sell occurs and we were at our most vulnerable after being supplied with copious amounts of free booze.

Once again, we were redirected to another part of the complex. This time, it was to a table and comfy chairs overlooking the main gardens and pool area. It was here that Helga launched into the question of financial arrangements assuming we were both sold on the idea and that she had done all that was required of her. At this point, I fully expected Iris to say that she was not interested and, that what we had seen was very nice, or something akin to 'it is not suitable for our purposes at the current time'. Unfortunately, for me, she did not and actually seemed to be quite keen to enter into a transaction to purchase at least one week of holiday ownership at this prestigious location.

I immediately intervened and said that I needed more time to think about it. This comment was met with a tirade of rhetoric and an accompanying sales patter from Helga that was to some extent supported by my wife. I decided to become a tad intransigent and put up a wall of excuses including the fact that I did not have that long to live (well another 40 years is not long in my book) and could not possibly take on such a venture. Both women seemed to gang up on me and Helga came up with the theory that if I died, I could bequeath it to my descendants (in perpetuity) and would therefore be an excellent investment.

I took a long hard swallow of my free wine and finished it off before slamming the glass down in an act of sheer defiance to show that I was livid. Helga, who had been sitting near to me with the sun at her side, caused a long shadow to fall across the table which somehow managed to silhouette her amazing figure and show up the outline of her bosom right at the point where my hands were resting. Then, because it seemed as though my hands were touching her breasts, this distracted me momentarily, to say the least. I was once again under her spell and ready to succumb to the pressure to purchase. Iris obviously noticed my fingers edging across the table top towards the shadow and she gave me a swift kick on the ankle which once again brought me back to reality.

I immediately reacted by standing up and looking Helga straight in the eye (I think it was her left one) and said, "No thank you, we do not want to buy. Can we please have our free gift of the jeep safari tickets?" I expected her to say 'Gehen hin Holle' (which roughly translates as 'Go to Hell') but she just came back at me with more sales pitch and even brought her boss into the proceedings. I must admit that they tried their best but this time around my stout resilience won through and they finally accepted defeat. They did give us our free tickets but refused us the customary lift so we had to walk back to our hotel.

I was totally relieved to get out of that situation and actually swore at Mrs J for getting us involved in the first place. She merely laughed and reminded me that we still had the free tickets and that it had been worth the hassle but I failed to see the funny side of the whole experience. The only redeeming factor was that I was probably one of a very small number of men who had actually touched the shadow of Helga's boobs and lived to tell the tale.

The long slog uphill to our hotel was eventually accomplished and we got changed to have a swim in the pool. This certainly cooled my ardour particularly as they had switched the heating off. The rest of the day was spent drinking and discussing what we should do the following day. That night I had a broken sleep with dreams of what might have been if Helga's shadow had been more accommodating and met up with my own shadow. The mind is a funny thing, you know, when affected by too much sun, too much drink and too much sales pressure. It can often lead you into a parallel universe where anything can happen. Well, let me tell you, mine does!

At breakfast the following day, Iris mentioned that she really fancied going on a Levada walk up in the mountains. I had no hesitation in agreeing to this proposal and I think that this unnerved her a little. She had become accustomed to having to fight for everything and a simple submission completely threw her off balance.

Chapter 9
Jeep-Ers Creepers

After a leisurely breakfast, we visited the tour agent at the hotel and booked one of the many walks available. The decision was based upon the level of difficulty and the length and duration which in our case was medium, half day and in the morning. It just so happened that the booking was to commence the following day and this left little time for me to brush up on my hiking skills. Sometime earlier in England Mrs J and I decided that we needed to take more exercise and in pursuit of 'getting things right the first time', we had gone out and purchased some suitable gear (or so we thought) and started engaging in the hobby of rambling.

Now 'rambling' as 'Doctor Johnson' defined it, is "An aimless peregrination of a rural nature taken by men who are tired of the town yet do not know enough of the country to hate it." I presume women are excluded from this quote because they never get tired of the town using it to remedy their ills via retail therapy. Well, to get down to the nitty-gritty, my first attempts ended up with an assortment of bodily ailments such as a sweat rash, foot blisters, a bad back, a head cold, a soiled walking boot (covered in cow dung) and a wasp sting. Add to this that I also nearly asphyxiated myself with the excessive use of midge repellent.

This, as you may gather led me to an early cessation of rambling and a stockpile of fairly expensive equipment which had to be left at home due to the weight restrictions on our flight. And so it transpired, that I was undertaking a mountain hike with little or no walking experience and only holiday clothing which, believe it or not, included a jumper. Iris on the other hand was a very adept walker, cyclist, aerobics enthusiast, healthy eater and a fast knitter. With this level of opposition, I was really up against it but decided to give it my best shot in true 'commando' style. That should really be translated to a 'come and do' attitude set against my normal exercise routine that was tantalisingly labelled 'SAS', which really meant 'swig and swell'.

When the time came, and our transport arrived, we had kitted ourselves out in jeans, sweatshirts, Jumpers, trainers as well as a waterproof coat for good measure. We had been told that the mountain air could be a little fresh particularly, in the mornings. To our dismay, all of the other passengers on our bus were kitted out in professional-looking hiking gear and appeared to have come to Madeira with the principal intention of doing just that. We reluctantly resigned ourselves to making the most of what we had and settled back for the journey into the region where the Levada was located. On route, we became friendly with a middle-aged couple called Alan and Marjorie who had, they informed us, been to the island on two previous occasions and made the best use of the walking trails that Madeira had become famous for. This particular walk was just a warm-up for the big one at the end of their holiday.

When we eventually reached our destination in the mountains, we were told to wait outside our minibus for another van that was bringing our packed lunches and hiking sticks. These latter items (we later learnt) were hand-carved sticks of differing lengths although each had a common feature of a thumb notch at the top end. As soon as the van arrived, I was like a school kid all over again and quickly made a grab for what looked to me like the best stick of the bunch. Mrs J dropped her head in sheer dismay at my actions and then shook it from side to side before looking up again and staring at me with a gaze that would melt carbon. This was something that I had become accustomed to on my travels with her and therefore the experience left me undaunted. I slowly moved away from the sticks to let the rest of the hikers have their pick.

Our guide for that morning had travelled in the van, with the lunches and walking sticks, introduced herself as Monica and that she was a local girl. This surprised me as she was fairly good-looking compared to the flower sellers and fishwives in Funchal. Her name, she explained, was common for girls on the island which surprised me as I imagined they would be called Consuela or Margarita, or even Sardine-ia in line with their Portuguese ancestry. She was a lively young woman who looked to be as fit as a butcher's dog and I am sure was more than capable of outwalking even the most experienced members of our small group.

We were all given the suitable safety talk about the pitfalls of the Levada's and after enquiring as to our relative levels of walking we set forth along our mountain trail towards our first rendezvous point which was a waterfall that I cannot recall the name of. At first, the going was easy but it soon became more precipitous and slippery which called for greater use of our trusty sticks. We passed by some beautiful scenery but there was little time to absorb the pleasures because Monica had set a blistering pace that allowed no time for anything other than rapid glances at our next foot holds.

When Iris and I did arrive at the waterfall, I was feeling very chuffed with myself because I had arrived second after our guide and, whilst we were waiting for the others to catch up, I managed to have a chat with Monica to learn more about the island and its highlights. I happened to mention the jeep safari we had acquired and after a sudden shocked look on her face and a small gasp, I realised that this perhaps was not going to be the experience we were hoping for. Anyway, more of that later.

After the stragglers caught up, we set forth yet again to locate the start of the Levada footpath. I use the word footpath very loosely because it really only suited the feet of a very agile mountain goat. The non-agile ones had obviously fallen to their death down the sheer drop that existed on the exposed side next to the irrigation channel which hugged the mountain side. It was at this point that chivalry and gentlemanly conduct prevailed and I let Mrs J go ahead of me as she had smaller feet and also that I could hang onto her. Well, you see this is another plus point of taking holidays with a partner.

After the trek, which seemed to take forever, we eventually reached the highlight of the walk namely the 'Pool of the Seven Springs'. Here we rested and availed ourselves of the packed lunch which consisted of a variety of sandwiches, cakes, fruit and a soft drink (one of those little cartons with a plastic straw you poke in into a designated hole). It was by no means fine food but some of it filled a hole (in a nearby rock crevice to be exact).

After our break, we were all called to attention by Monica and escorted camel train fashion along another part of the trail. This led us towards, and through, a dark and very wet tunnel before exiting out into the wilderness once again and onwards towards our eventual pick-up point further down the mountainside. This walk back was no less easy and required dextrous use of the hiking stick to maintain a vertical stance. Eventually, we reached base camp where we all boarded our minibus for the return journey to our respective hotels.

The next day was spent around the pool and walking to Funchal to explore the city a little better with its many attractions. As for the jeep safari, we had planned it for the last full day of our holiday, and when the great day came, I had cause to remember the look on Monica's face when I mentioned it to her. Undaunted we set off to the designated pick-up point and awaited the arrival of our transport. When it finally appeared, some fifteen minutes late, I was surprised to see that it was not the kind of jeep I had imagined. Instead, it was one of those all-terrain vehicles built in some communist country and was basic, to say the least.

Our fellow passengers consisted of a 'well-built' Glaswegian man and his father as well as a family of Germans who did not speak any English. This was going to be interesting, I thought to myself, especially if they were related to Helga. Our driver for that day was called Fernando and was a strange character who did his best to get to know us all but (as you can imagine) the odds were stacked against him as he spoke only a few words in English and none in German.

We eventually, with the help of some sign language, arrived at a few basic terms such as 'Quack, quack' when we approached an overhanging tree branch and 'PP open stop' if you needed a comfort break. These roughly translated to 'Duck' (your head) and 'Forest Toilet' (if you were bursting for a pee). The journey commenced with a passage along a rugged mountain road until we met up with another similar vehicle. This was when Fernando announced that the fun had, or was, about to begin.

We were told that standing up in the rear of the vehicle was preferred to sitting and as soon as we hit our first rock, in the road, we realised just what the driver meant. The German family had stood up because everyone else had but had no idea why until that first bump. This brought about two things. Firstly, it generated a rash of white knuckles on all of the occupants as we hung onto whatever we could find and secondly, we quickly learnt to 'ben-ze-kneeze' which the German family were good at probably because they were experienced skiers.

The track got progressively more rugged and it took all our wits to stay upright, learn to drop our heads on the command of 'Quack, quack' from the driver, and above all try not to fart. The latter was the most difficult part because Fernando was hell-bent on knocking the s**t out of us. I had decided to hold Iris with one arm and cling tightly to the roll bar that was directly behind me. Iris however decided to grab hold of the rollback tarpaulin roof cover that was tied behind the driver's seat by means of two leather straps. She managed somehow to pull it from its rivets and caused it to produce a sickening rip-like sound that luckily Fernando failed to hear.

As this strategy had failed miserably, she next decided to grip my arm so tightly that the circulation stopped and my arm went numb. As my arm turned purple in the afternoon sun, my thoughts turned to self-survival with the result that I failed to hear the next 'duck' command and received a thwack from a tree branch as a punishment. My knees were buckling under the strain of holding Mrs J upright and it was at that precise point that I realised why they gave these jeep rides as a free gift for not buying holiday ownership. My thoughts turned to Helga (who by now had taken on the persona of the grand-daughter of a prominent Nazi) and this was her retribution for me having had the audacity to touch the shadow of her left breast during a momentary lapse of concentration at the sales venue.

Another smack across the face by a low-lying branch caused me to utter the cry 'PP Open Stop' as a means of stopping the jeep. I didn't really want a comfort break but I had reached the end of my tether and needed to go off into the surrounding foliage to have a good cry without anyone seeing me, especially the German family. After my feigned sortie into the woods, I returned to join the group only to find them attempting to identify some herb growing wild at the side of the track. I had been to hell and back and they were all trying to identify a bloody plant.

As no one had a clue, I thought that I would throw in my ten-pence worth and announce that it was probably 'Oregano'. Everyone went silent awaiting Fernando's response. He paused before declaring (in broken English) that I was a 'clever dickie' and had got it right. I did not really know what it was but shouted out 'oregano' because it was the only one, I could think of at the time (or should that have been ‘Thyme’ which I thought of shortly afterwards). The husband and father of the German family finally uttered out loud, “Yah, es gut. Oregano. Ein kraut,” or something similar. The sly monkey had kept it a secret that he actually spoke.

After this enforced stop, we all re-boarded the death wagon and sped off along the track which was mostly downhill after that point. After several ‘Quack quacks’ and a couple of new quotes that we all failed to grasp the significance of, we eventually arrived at a small hillside village where we were able to alight and take advantage of a stop for refreshments. This was most welcome after our recent experiences and some fifteen minutes later we set off again but this time we had re-joined proper tarmac roads and could once again sit down in the back of the vehicle.

We soon reached our designated drop-off point and subsequently returned to our respective hotels. At ours, we immediately made for the bar and availed ourselves of some much-needed alcohol to settle our nerves before getting changed to enjoy dinner. That night was spent reliving our experiences of the holiday before we succumbed to sleep.

The next day we had to pack in readiness for our return flight but still managed a final visit to Funchal to once again take in the sights and sounds of this vibrant place. The end of the holiday seemed to creep up on us rather hurriedly and as we bid farewell to our newfound friends, at the hotel, we could not help feeling a tinge of sadness to be leaving such a wonderful place. We decided there and then that we would return again one day in the not-too-distant future armed with a much greater knowledge of the experiences that this island has to offer.

I am pleased to report that the return flight take-off was much less hazardous and that we arrived on time at Manchester. On this rare occasion (I am further pleased to report), we had booked an additional one night stay at the same hotel, where we had parked the car. We did not fancy a drive home late at night as we had done in the past.

Chapter 10
The Flight of Flights

With a misty recollection of our previous soiree to the Caribbean still in our mindset, we set forth to the travel agents to book another vacation. This time however it was to be to the true land of 'No Problem' namely Jamaica. This vacation opportunity came up by chance and as the price was good value, we decided to book it even though it was once again from Manchester Airport in March 2001. As before we managed to acquire overnight accommodation at a hotel that provided free car parking for the whole two weeks of our holiday. On this occasion, however, there was the added bonus of the use of spa facilities (at the hotel's leisure club) and a free shuttle service to and from the airport. It was a no-brainer seeing as the total cost was cheaper than parking alone would have been.

The next day was both damp and dismal with a faint dusting of snow. The dreary scene failed to alter our high spirits as we proceeded to dress in lightweight holiday clothing in order that we would both arrive in style and not have to suffer the heat in our normal travel clothing. This was a new approach by us gained from our previous travel experience to the Caribbean and we were prepared on this occasion to suffer mild hyperthermia until we reached terminal one.

We hurriedly consumed our own pre-packed breakfast, which included two ageing semi-soggy bananas, before securing our bulging suitcases. Iris has always been keen to pack for every occasion and known weather condition, even though this brought us dangerously close to our baggage allowance. This trip was not going to be an exception and I recall her packing at least ten different swimming costumes, some of which were for photographic purposes only.

At the allotted time, we presented ourselves at the front of the hotel awaiting the arrival of our airport shuttle, which was in fact a free taxi service. My thoughts wandered slightly at this point wondering how it was that the hotel could offer such cheap room rates, free parking and a free return taxi ride to and from the airport. Surely, they would go bankrupt with these concessions. Worry quickly overtook these thoughts when our designated taxi failed to materialise and I had to request a rather sullen-looking hotel porter, on reception duty, to ring the firm and enquire as to the delay.

The porter made some enquiries and after a short while notified us that they were pulling up outside as we spoke. I quickly returned to my lookout position, in the snow-covered entrance way, just as a very sporty BMW coupe pulled alongside. In amazement (and disbelief) at our good fortune, I rapidly gathered together our luggage into a well-organised grouping and prepared to open the rear door. At this point, the BMW drove off leaving me somewhat bewildered and a little embarrassed when I realised that it was not our taxi. Heaven knows what the driver thought of me particularly dressed in lightweight summer clothing.

A few minutes later, a rusty old Toyota pulled up and the driver, an Iranian by descent, mumbled some incoherent sentence that had the word 'airport' in it, so we assumed it was our shuttle service to terminal one. He threw our luggage into the boot whilst we slid into the rear seating area of the car. All windows, with the exception of the windscreen, were badly frosted up to the point where it was impossible to see out of them clearly. He was, in my estimation, driving purely through forward vision together with a wound down driver's door window allowing side vision to the right only and a minimal rear view with the assistance of a cracked wing mirror. The last thing we wanted at this time was air conditioning 'au natural' especially as we were dressed in nothing of substance.

The driver was frustratingly slow, far too slow for my stressful condition. I have a weakness for being on time and that includes airports. I live in the forlorn hope of one day being able to request seats with extra legroom and actually being granted them without having to pay a premium for the privilege. This obsession, which involves getting to an airport three or four hours in advance, generally comes to no avail but I continually live in hope. On this occasion, we were only 2 hours and 50 minutes early and I had decided that war with Iran was the only solution to console myself with. The driver foolishly expected some form of a tip, for extracting our luggage from the boot, but I simply grinned at him and we rushed off to the ticket counter before I did something that I would later diplomatically regret.

As luck would have it, we managed to secure a double seat at the rear of the aircraft that we later learnt was a DC10 of vintage pedigree but whose overall size was impressive.

The allocated seating position was also to provide us with other bonuses such as additional leg room, good access to the toilets and the chance to be served first with food and drinks. For some unexplainable reason, I started to feel uneasy at our good fortune and even sensed that this could end up being our last ever flight and the great god of aviation was sending me off in style. These fears were climaxed when a voice of the airport tannoid system announced that "Airtours Flight A1210 to Montego Bay, Jamaica was to be delayed by up to three hours due to a technical problem." This problem I learnt later was a hydraulic brake failure which the captain had detected upon his inspection of the aircraft. Accordingly, he had ordered his entire crew off the plane until the ground staff and mechanics had made good the repair.

Ah well, I thought. Better to have discovered it before take-off, than in mid-air or just before landing. Believe it or not, this was the first delay we had experienced on all our travels by air. The extended airport time was difficult to handle as it was too late to return to the main shopping and refreshment areas and our boarding gate area had very few facilities other than a toilet. In order to reduce the boredom, I decided to commence reading my newly purchased book on *How to write poetry* that I had acquired with the intention of devising a sonnet or two that I would dedicate to my wife for our Pearl Anniversary that was going to be celebrated during our holiday. Thirty years of marriage, to the same woman, would surely produce some meaningful rhyme that I could be proud of and that she would be impressed with. The trouble is that when you start to learn the mechanics of writing good prose it suddenly becomes extremely difficult and removes the spontaneity of thought. The best that my pencil and anxious mind could muster at that precise time was the following ditty;

Wood and water, sun-bleached sand,
Jamaica is a tranquil land.
Of happy people proud to be,
A culture of diversity.

Don't give up the day job as they often say, but I was only at chapter one in my poetry book and after all it helped pass away some of the waiting time. This was merely a warm-up for the great modern sonnet of fourteen regular lines, a double stanza and the obligatory 'ABBAABBACDCDCD' pattern of rhyme. Oh! What it is to be gifted.

A short doze ate up the remaining time until the all-important call for boarding was finally announced. Excitement overtook any faint doubts as we entered a DC10 for the first time. I was immediately struck by the Mary Quant décor and mock William Morris curtains over the windows. These gave the plane that quaint sixties feel which momentarily took me back to my teenage years. For all I knew, many famous people could have flown on this aircraft, icons like Glenn Millar, Buddy Holly and Jim Reeves to name but a few. Nah! Only kidding. *Never mind* I thought, *if it is good enough for 'Airtours' then it's good enough for us*.

Despite this being a really old aircraft, Iris and I soon found ourselves settled in our seat at the rear. Jokingly I asked her to have a look through the window (to check how many sets of wings it had) to make sure that it was not a bi-plane. After having a quick glance through the 'in-flight' magazine, I did what I normally do prior to take-off and that was to weigh up the other passengers and check over the cabin crew. The next job was to read the emergency card and its details should a crash prove inevitable. One thing puzzled me greatly as I could not see any point where oxygen masks could be released from their normal position below the luggage lockers above us. Eventually, we discovered that they were concealed in a small section above the fold-down meal trays on the seats in front of us. This was of great comfort in that the plastic cover of the masks would probably smash into our foreheads and render us unconscious before we had any chance to use the masks.

These thoughts were quickly dispersed as the engines burst into life and the aircraft started its slow journey to its designated point for take-off. Before very long, we were airborne and flying over the Irish Sea bound for the coast of America. The captain had informed us that he was flying west to Boston, USA, for re-fuelling, before turning left down the east coast to Jamaica. He further stated that our anticipated arrival time would be 18.30 hours local time. The adventure had begun.

As I lay back in my, decent legroom, seat I thought to myself, *That's it. Nothing else is going to spoil my holiday*. Think positive as I have often been told. Wrong! Again!

Halfway across the Atlantic, two separate domestic arguments broke out. One involving a West Indian family, travelling to see relatives back home for a wedding and the other between a white British couple who, as it turned out would have been better placed to have taken separate holidays. The respective rows got quite heated at times and we thought we were about to witness our first round of 'air-rage' but the start of the in-flight movie ('Billy Elliot') seemed to smooth the situations and calm returned to the cabin for a while. A 'tannoy' announcement from the captain broke the silence and disclosed two things. Firstly, that we had reached the eastern shores of the USA and the second was in regard to the fact that someone had been detected smoking in one of the rear toilet compartments and he emphasised that this was a no-smoking flight.

Naughty, naughty whoever it was. Well, we both knew actually because he had just recently passed our seats, en-route to his own, and he smelled strongly of tobacco smoke. *Should I grass him up?* I thought, and claim the reward or simply ignore the whole incident. I chose the latter seeing as he was a six foot, plus Jamaican who resembled the leader of a drugs cartel that I had seen recently in a television documentary. "Stay cool, man," I kept telling myself. "Chill. You are on holiday."

Suddenly pure mayhem broke out and air crew were dashing up and down the plane asking if any of the passengers was a doctor, a nurse or someone who knew anything about medical matters. It transpired that one of the passengers (at the front of the cabin) had collapsed with a suspected heart attack and no one had a clue how to deal with the situation. Believe it or not, two other people also collapsed with similar or separate conditions and as such the captain announced that he may have to divert to an American airport to transfer the ailing persons to the nearest hospital.

Even though I was a serving police officer, who had been trained to deliver babies on the top deck of buses and who had a bronze life-saving award (earned at the deep end of a swimming pool), I felt my expertise did not stretch to such mundane complaints and as such declined the request for assistance. As luck would have it, all three passengers recovered after a glass of 'Airtours' tap water and accordingly we continued on our chosen flight path. Obviously, this water had some special healing powers probably due to the fact that the brain tells the body to heal itself quickly to avoid having to consume any more of the stuff.

The flight eventually ended and the plane taxied towards its allotted parking zone at Montego Bay Airport. The doors opened and a sudden gush of extremely hot air engulfed us as we filed off in an orderly fashion. "This is more like it," I uttered to Mrs J especially as we were kitted out in our summer clothes and ready to embrace our Caribbean holiday in true Jamaican style. The big drawback was that it was 6.30 pm Jamaican time and 11.30 pm British time which was way past my normal bedtime. Still, we were not going to let this mere triviality ruin it and prepared ourselves for the rigours of the Jamaican customs and immigration services.

Unfortunately, they were determined to scrutinise every passenger due to an outbreak of foot and mouth disease back home. As there were some 350 passengers, this took some considerable time which further delayed our onward journey to our chosen hotel. Despite my efforts to remain 'cool', I was getting tired, irritable and very hot and I could sense Iris was feeling it as well. The temperature was in the eighties and it felt very humid. However, once we were clear of the formalities, we were met by our representative and duly directed to a waiting coach in a large car park at the furthest possible point away from the terminal building.

One of our suitcases was whisked out of my hands by a local porter and before I could say, "No, thank you," he was off at a gallop towards the waiting bus. I did my best to catch him but failed as I had the heavier of the two cases. He seemed very pleased with himself that he had beaten me to the coach and, with a big smile and a set of white teeth to die for, he asked me If I would like to show my appreciation. I started to applaud his efforts (as appreciation) but he seemed to get angry and told me he wanted a tip. *Cheeky bugger,* I thought but decided to give him $1 to put an end to his stupid grin. By this time, I was in a fighting mood and dare say that I could have given Mike Tyson at least 20 seconds. Tipping it seemed was to become a way of life.

We finally arrived at our hotel which was called 'Club Ambience' that was located near a small cove known as 'Runaway Bay'. We were weary, sweating and ready for bed but not before sampling the delights of a very strong rum punch that went down a treat. Our room was very tasteful and contained one of the biggest king size beds we had ever seen. It was unusually comfortable and resultantly were both soon off into the land of nod. Jet lag! What jet lag?

The following morning, I was the first to wake up. Probably due to my body floating in a new time zone, or more likely because of a cacophony of differing background sounds that included several cockerels, a convoy of heavy goods vehicles (passing on the nearby road), the hum of the air-conditioning unit and the hotel staff shouting at each other in a strange language. This language is known as 'patios' (pronounced 'patwah') and is a mix of English, Spanish and French but is learnt on the streets rather than being taught at school. As it turned out, we were to be given lessons in it later in the holiday but we failed to master the dialect because of the high speed at which it is spoken.

Whatever the reasons, I was now wide awake and ready to discover our new surroundings. I was not prepared however for what met my gaze as I peered out of the curtains before venturing out onto our private veranda. The weather was gloriously hot, we had a garden of banana trees and succulents and this opened up onto a turquoise blue swimming pool with the beach and the Caribbean Sea beckoning just beyond it. It was, to say the least, idyllic and I quickly remembered why it is that we, and so many other people, adore this region of the world. I walked around for a while getting my bearings before heading back into our room to wake Iris as I knew she would not want to waste a minute of the weather or the ambience of the place. Hence the hotel's name.

After a wonderful breakfast which included the best cup of coffee I have had in my entire life, we attended the obligatory representatives group meeting to hear about the rules of the hotel and the many optional excursions that were available at set prices. The package we were on was all-inclusive which meant that everything, including alcoholic drinks, were free and I was determined to ensure that I did my best to keep the flag flying for GB and not let any visitors, especially from the US of A, gain an unequal supply of the available rum and other high-octane delights. Such was our new lifestyle and an air of intoxicated calm became the norm for the next few days of our stay. We had little cause to explore the island as there were so many activities and other vices to entertain us and besides there was always the second week when boredom usually set in. The plan worked fine and before we knew it, we were part of the established residents following the departure of the previous weeks holiday guests.

The ambience that the hotel used in its name, and from which it gained its reputation, was to be sorely put to the test following the arrival of a twelve strong wedding party from Monday's Gatwick flight. They were a mixed bunch intent on having fun and they wasted no time in establishing themselves with week old veterans like Mrs J and myself. No one was left in any doubt that the hotel was to be another watering hole made more welcoming by the unending supply of free booze.

Chapter 11
High on the Vodka, Low on the Coke

One of this new bunch was an affable young man, fast approaching thirty, called Dale. This was in fact meant to be his middle name but he claimed that his father had mixed them around at his christening and Dale (who was only one year old at the time) was in no position to argue about it. So 'Dale' it was and on hindsight, I would say that it suited him and matched his outgoing exuberance for fun. I presumed he was gay because he talked with a very camp voice (that was a cross between 'Julian Clary' and 'Alan Carr') although outwardly, he favoured the company of females regardless of age, ilk, pedigree or background and was prepared to speak to anyone, even me. He was thinning a little in the head hair department and was clearly attempting to develop a physique not too dissimilar to my own. I had a few years advantage over Dale and had always regarded my body as a temple to neglect. Clearly, he had some way to go but was nevertheless into some serious weight gain training to emulate the pot belly, I sported, that was highly sought after by so many men.

To these ends, he became synonymous (amongst our group) with a hitherto unrequested mixed drink at the hotel. This he self-styled 'high on the vodka, low on the coke.' This seemingly simple request offered a direct challenge to the grumpy pool bar attendant and forced him to re-think his strategy in the use of his bottle top measuring device. In fact, it forced him to abandon it completely in desperation. Dale's 'camp' and very persuasive manner clearly rattled him but he tried hard not to show it for fear of being demoted to the outpost bar at the end of the hotel's own pier. This was obviously the most feared of all the hotels staffing assignments.

Club Ambience maintained the dictum that 'the customer is rarely ever wrong' and appeared to send all their staff, with the exception of the grumpy pool bar attendant, to the Disney School for Forced Smiling. Either that or the staff were proud of their teeth which glowed bright white against their black faces every time they had cause to smile which seemed to be quite often. And so, with the passage of time, approximately two hours of constant requests, the new cocktail (high on the vodka, low on the coke) was indelibly etched into the barman's repertoire and he managed to get the proportions spot on every time.

Life around the pool became a total hoot and the humour seemed endless. We even had a visit from a water aerobics teacher who resembled Grace Jones both in looks and physique (totally scary). She was however very likeable and soon had all the women in the pool gyrating their pelvises to the points of the compass in an east, west, south then north movement. This was to the vocals of 'five cents' (east), 'ten cents' (west), 'fifty cents' (south), 'a dollar' (north).

However, she made everyone thrust their hips forward on the final push north. There was a chorus of 5 cents, 10 cents, fifty cents, a dollar! which was hilarious to watch and was causing quite a stir in the pool as all the men were cheering the ladies on especially with that final forward thrust.

Life continued in a semi-drunken haze until the start of our second week when we decided to book a couple of excursions to see a bit more of the island. One was to the famous Dunn's River Falls near a place called 'Ocho Rios' and the other was a bamboo raft trip along a river called the Martha Brae. This latter trip was near to another north coast resort named 'Falmouth'.

Dunn's River Falls is a 'must do' for all visitors to Jamaica and even the locals themselves regard it as a unique experience and go as frequently as they can afford to. On our visit there, we had the assistance of a guide called Daniel whose smile was a wide as the falls themselves. We were, on this occasion, teamed with an American family of five in total and Daniel had us all linked in a human chain to climb the best part of the 100-foot plus ascent which commenced at beach level and upward to the main exit point. We were left in no uncertain terms that the trip was a total wet experience and had changed accordingly prior to starting. It turned out that the wet experience was a total understatement as by the second photo shoot (Daniel had our camera), we were drenched to the bone in fresh mountain water which was cool but not cold. The climb, however (once completed), was a unique memory and the suns heat quickly dried our lightweight clothing.

When we had exited the falls, Daniel asked for a tip, in appreciation of his guided tour, and as I only had a $20 US bill, I asked if he had any change. He swore blind that the smallest note he had was a $10 bill and therefore his tip sadly amounted to 10 dollars. After securing that amount from us, he turned his attention to the American family (husband and wife plus three kids) who had the embarrassment of informing him that they had brought no money with them. Daniel's face nearly turned white and the look he gave them was tantamount to a killer curse. At seeing this response, the father of the group stated that he was willing to give our shuttle bus driver the tip when he got dropped off at his hotel and then the driver could give the tip to Daniel the next time he saw him.

Daniel simply walked off in disgust and I thought, to myself, that this cunning ruse had worked and that they had got away with it. Why didn't I think of that ploy? We then all boarded our transport and set off to the first hotel where we dropped off the American family. To our utter amazement, there waiting at the hotel entrance was Daniel holding out his big hand waiting for the promised tip. He had literally driven about two miles in his own car to obtain a similar tip to the one we had given him. This tipping game was clearly big business in Jamaica and Daniel was one of its professional exponents.

The second excursion to the Martha Brae River was less fraught although tipping was still expected. I, however, was more prepared on that occasion and made sure that I carried a number of smaller denomination notes. The river ride, which was taken on elongated bamboo poles fastened together to form a raft, was also unique and afforded us a magical ride down what seemed like a jungle river. Our raft captain was called Wade and he gave us the benefit of his vast knowledge about the wonders of the Jamaican rain forest as we glided along the peaceful flowing river. The downside was that it was searingly hot and there was no protection from the sun or insects. The river's name is a corrupted version of 'Rio Mateberino' and there is a legend attached to it. The story goes that a 'Taino Witch' was tortured by Spanish settlers into revealing the location of a gold stash hidden along the river. After divulging, she changed the course of the water, killing the Spanish and blocking the cave where the treasure remains hidden. Ah well, there you go, and now you know. (Woah! Another spot of poetry creeping in).

I have to say that both of these excursions were a wonderful experience and we were glad that we had made the effort to go as it added another dimension to our holiday. After we made it back to our hotel, we once again resumed our shenanigans with our friends enjoying the ambience that the pool, sunshine and free drinks offered. To all intents and purposes, life was serene.

That was of course until the arrival on the scene of 'Ritz'.

Ritz was a German tourist (in his mid-sixties) who, although married, had a style not too dissimilar to an ageing pervert and he was clearly on the lookout for any stray females younger than his wife who I would estimate was of a similar age to his own. As single females were in short supply, he turned his attention towards the married women, including my wife Iris and he seemed unable to avert his lecherous gaze whenever she passed by him or was seated near to him.

He eventually plucked up the courage to come across and talk to us and our newly established friends Eve and Iris (Yes! another one). You can go through life never meeting an 'Iris' and suddenly another one comes along just like buses. Being in the company of two at the same time was mind-numbing.

Anyway, it was at this point that I realised what an apt name Ritz had because he was undoubtedly 'crackers' or at least one banana short of a bunch. That cannot be tolerated, particularly in Jamaica where the people pride themselves on full bunches.

His command of the English language left lots to be desired and he started rambling on about how Germany was now free of what sounded to us like 'rabbisch'. Eve quickly retorted that she wasn't impressed due to the fact that we now had 'foot and mouth' and it was spreading all over the place. He, and the rest of us, failed to grasp the significance of this statement and as we were doing our best to ignore him in a true British 'stand-off', he eventually wandered off in a rolling gait that gave measure to the amount of alcohol his body must have consumed that day. At this point, I mentioned to Eve that I believed he meant that Germany had rid itself of 'Rabbi's' and this gave way too much laughter as she thought he had meant 'rabies.'

Eve and Iris (the other one) were holidaying together for the first time. Both had recently been widowed from their husbands and had met due to the fact that they were both market traders in the Lincolnshire area of England. They made excellent companions and both Mrs J and myself enjoyed their light-hearted banter for the remainder of our holiday.

On the following day, the four of us had been conversing around the pool area when Ritz appeared and made a bee-line for our group. He was wanting to go to the local church on Sunday and was building it up as some Broadway production with singing and gospel readings. Religion was apparently another pastime of Ritz's and he expected everyone else to be as keen as he was. During the conversation, he switched to giving advice about the Euro and the purchase of shares on the American stock market. It was difficult at the best of times to understand his broken English and the rapid switching of the subject matter was not helping in the slightest. Thankfully, he departed for a short while when Dale appeared on the scene. We duly warned Dale about Ritz which was just as well as he suddenly re-appeared, without warning, almost like a 'Stuka' bomber attacking out of the sun.

On this occasion, he turned his attention on Dale and did his best to persuade him to go to church with him. This was a bad mistake on Ritz's part as Dale was not into religion and when he failed to convert him, he attempted to exorcise the demons that he thought were possessing Dale and therefore clouding his judgement. I swear, at that moment, that I was about to witness the first recorded incident of 'pool side rage' at the 'Club Ambience' and decided to back away from what was about to happen to Ritz. However, as luck would have it, Dale was in a benevolent mood and told Ritz to back off or else he would 'Let the Dogs Out'. Ritz did not fully understand this comment but nevertheless took it as a serious threat and wandered off to join his wife. By some strange quirk of fate, we did not get any further trouble from Ritz so life around the pool returned to normal.

During our penultimate night (at the hotel), we retired to the lounge area after the evening's entertainment where we once again teamed up with Eve and Iris for a nightcap and a chat before retiring to our room to get a good night's sleep. I have vivid recollections of our conversation because the subject was one of bleach. Yes Bleach! Now, although I know a little about the product, I was amazed as to what was discussed. There we were, the four of us, discussing the why's and where-fore's of bleach on a balmy night in Jamaica and going into some detail about quality, colour, germ control, smell and overall effectiveness. It seems that they put the stuff to many more uses in Lincolnshire than in most other English counties. And no decent home would be without it. I actually became quite over-reactive at one point as I wanted to make, what I thought to be, a very relevant point about the use of rubber gloves. I was told to calm down by Mrs J as she claimed I was being a diva and that I was trying to divert the subject matter. I then departed and went to our room as I realised the drink was getting to me.

When our final day arrived, we did our packing and then bid all our newfound friends farewell before enduring the, one and a half hour, bus ride back to the airport. The roads and the toilet paper were the only two negatives of our holiday. As they say *all good things must come to an end* and so it was with this wonderful holiday, as we touched down in Manchester faced with jet lag and a three-hour drive home.

Hang on a minute! What about the sonnet I promised you?

Well, I did muster one attempt during and shortly after a good drinking session. The pen was willing but the mind was not at its best. However, in the interests of fairness and to avoid any broken promises here it is.

Sonnet

A passing thought the mind once chose,

Some moment lapsed borne out by dreams;

May yet transpose in foretold schemes,

And manifest itself in prose.

What sentenced powers lone words may lose,

Gain strength in kind from skilful pen;

To summon quotes, embellish them,

Or increase by synonymic clues.

Yet frail echoes reverb from soft phrase,

A thesaurus may provide the choice.

In temperate measure much to praise,

In verse or song much to rejoice.

To capture thus in many ways,

And make possible by human voice.

(Tony James)

Well, I was only up to chapter three in my book! Clearly, I have far to go and intend to if time permits.

Chapter 12
www.cancun@mexi.co

When the time came for us to make the decision for another holiday, our choice was somewhere in Mexico as it sounded rather exotic and believe it or not the prices were not too bad. Alas, we forgot all about the required vaccinations and their aftermath. Thankfully, we only had to suffer one additional jab from those administered on our previous trip to Jamaica and this was the one for 'Typhoid'. Now you may wish to learn that the word 'Typhoid' is derived from the terms 'Typhoon' and 'Haemorrhoid' and it is usually given to alleviate the suffering of the latter whilst enduring the former. With this in mind, it was of great comfort to know that we had the necessary protection from such meteorological discomfort and with a full set of inoculations we now felt capable of tackling any hazard, even jogger's nipple in a drought. Ces't la vie.

My biggest worry was the expected high temperatures coupled to high humidity that prevails during the summer months from June to September. Sweat rash was not my favourite complaint and I made sure I packed the correct medication to deal with such issues. You can always spot a sweat rash sufferer by the way they walk. If it's not the wide-legged John Wayne swagger, then it's the tight-legged waddle as if the victim had an incontinence problem. One wonders how the locals deal with these little irritations (the complaint that is not the Mexicans). I made it my quest to find out the best way to survive in extreme conditions.

Our chosen destination in Mexico was Acapulco on the Pacific coastline as it sounded very exotic and posh. We accordingly selected our hotel and paid for the holiday. However, after booking, I did some research on the resort and to my dismay learnt that there were localised problems such as city smog, dangerous undersea currents and even dead dogs and rats being washed into sea, from dry river beds, after a heavy downpour of rain. These by the way were the highlights. On the plus side a friend of ours, who had visited Acapulco a few years before, told us that it rarely rains and that when he went it was so dry that he used to joke 'he had seen a tree chasing a dog'. The mind boggles! However we were totally put off by what we had learnt and decided to see if we could cancel the holiday at the travel agent's we had booked it through. The problem was that we still fancied a trip to Mexico because it was one that had remained elusive to us until that point in time.

When we visited the travel agents, they suggested we try a holiday on the opposite coast namely a resort called 'Cancun'. This was located on the Yucatan Peninsula and is classed as being adjacent to the Caribbean which was one of our favourite zones in the world. Cancun is in the state of Quintana Roo an area that is mostly flat and although largely covered in tropical rainforest it boasts some of the most beautiful white sandy beaches you will ever see. Here the sea is crystal clear and changes colour subtly throughout the day from pale aqua to a deep turquoise (at noon), to cerulean blue under the blazing sun. It then turns pink with a splash of mauve during sunset. I hope you are impressed because there is more to follow later on.

Having painted an idyllic picture of the place I must mention the irony in what we found after our arrival there. This was that the resort should have been called 'Can-Con' with its flashing dollar signs and its rip off prices The locals seem hell-bent on fleecing you of as much money as they can get. My local guide book, or should I say my local library's, does not tell you this but does give an interesting account of the regions origins which I would like to share with you. Apparently, in 1967, a large Mexican number crunching computer selected a small swampy piece of land surrounding an isolated lagoon. The computer, as if by magic, declared this area to be Mexico's most promising tourist destination and (there you have it) 'Cancun was born'.

Now that I have set the scene, on with the nitty-gritty facts of this holiday which had been taken to celebrate Iris's birthday, but the imminent threat of castration prevents me from revealing which one. We had once again flown from Manchester Airport only this occasion the plane was as modern as they come, namely an Airbus A330, which boasted the first lower deck toilets on any aircraft at that time. These however could only be used when cabin staff were on stair duty; a task we assumed that was reserved for the newest recruit or someone on disciplinary punishment.

The flight, despite being delayed, was largely uneventful and everything seemed to be going well until we reached Cancun International Airport. Before disembarking, the entire set of passengers were sprayed with a fine mist of some unknown substance presumably due to the foot and mouth outbreak in the UK during 2001. Further evidence of the decontamination process occurred prior to our retrieval of luggage which arrived in a damp condition on the baggage conveyor after a considerable delay. Many passengers (including us) complained but we were fobbed off by airport staff stating that it was due to the high humidity. Ha!

Once we had retrieved our suitcases, the queuing began for customs scrutiny and that turned out to be something of a lottery. Once we had reached the front of the queue (which seemed to take for ever), we were directed towards a set of traffic lights and requested to push a solitary button. If it triggered a green light, you could pass through unscathed but if it showed red (which, you guessed it, ours did), you were instructed to go across to a nearby table and have the waiting customs officials search your luggage. We had nothing to hide other than my sexy new swimming trunks that I feared might breach the local laws of common decency for an 'hombre'.

Believe it or not, the customs official spotted them and held them up for examination. I looked him in the eye (not sure which one) and pointed to my wife as if she was to blame. He looked back at me (with the same eye), smiled, and continued his search until he was satisfied that we were not concealing any arms, ammunition, drugs, currency, illegal immigrants, foodstuffs or any flora or fauna from Manchester. Mind you, I had not seen any in the city before our departure, apart from a one-winged pigeon that had been tagged with a leg ring, naming it as 'Lucky-No. 3492'.

Once the airport hurdle was cleared, we boarded our transfer coach and met up with our courier who was called Fernando (no hide-ing-away with that name). He spoke in a form of broken English and delighted all the passengers with his account of the region's archaeological highlights. His description of the famous Mayan temple 'Chitzen Itza' sounded more like 'Chicken Pizza' but we managed to grasp what he meant. He did his best to promote his luggage slave William in the hope of ensuring him a share of any tips upon our arrival at the hotel. Despite this, no one seemed interested in parting with any money particularly as we were all pretty fed up and tired by that time. We decided not to tip because we only had two cases and ours were considerably lighter than most.

Upon arrival at the hotel reception, we were given our statutory 'all-inclusive' coloured wristband. On this occasion, they were aquamarine blue in colour and a thought struck me that perhaps I should compile a reference book on the subject entitled 'All-Inclusive Wristbands of the World-A Spotters Guide'. Then it dawned on me that perhaps my liver would not endure all the research necessary to complete it.

We received our room key and managed to locate our room, for the duration, on the second floor. Once we had unpacked, the pace of life settled a little and we became aware of the high levels of humidity. I was leaking like a colander, despite the air conditioning, and even Mrs J was suffering. Now on the face of it, that does not sound like much of a statement until I mention to you that she is the type of person who thrives on heat, hates the cold and would normally need to wear a cardigan in a tramp steamers engine room. I was actually witnessing her perspiring and realised that she was in fact humanoid after all.

I had this sudden and desperate need to top up my hydration and so we headed directly to the bar to order a drink.

Any drink! The problem we encountered was that none of the bottle labels gave any indication as to what they contained so I had to ask the barman (who I later found out was called 'Asterio'), for assistance. Luckily, he could speak a smattering of English and suggested we try a Margarita. I told him it was liquid refreshment I was after not the local whore. He failed to grasp the gist of my attempted humour (probably because of my north-east accent) and shrugged his shoulders at my comment assuming that I was requesting a new cocktail that he did not know how to make. Well, I suppose I could have had a 'Bananwhore Diquari' or a 'Vodkwhore and Tonic'.

After this Mexican stand-off, we finally settled on a local rum and cola (as my mother had always warned me about drinking water in foreign lands). The drink tasted odd but was consumed all the same as we were so thirsty, and to be honest, I could have drunk gnats pee if it was available. I was in reality becoming so dehydrated that I was fire risk and so the soothing qualities of free booze (well you know how it is) were the only remedy worth taking. After two further drinks orders, we went in for dinner and then decided on an early night's sleep. This was mainly because we had been told not to miss the welcoming party at 9 am local time the next day.

The night passed quickly and we were both awake fairly early and raring to go. I kitted myself out in a T shirt, shorts and sandals together with my trusty shades for that all round macho look. We were early enough to have a leisurely breakfast before assembling at the designated room for the welcome party. When our rep arrived (at the appointed time), she turned out to be an English girl of small proportions with a round face that housed a gob capable of formula one linguistics. Never in my life have I heard anyone talk so fast and for so long without taking a breath.

I very quickly formed the opinion that she was an 'Airtour's android' programmed to deliver the same talk each week but somehow her fast-forward button had stuck and her lips became a blur. During the talk, we were apparently given advice on what to do, what not to do, and where we could venture for fun and frolics. My problem was that I failed to absorb most of her speech and was left to mull over the feeble handout, we had been given, in order to make our bookings for the excursions on offer. We actually chose the 'Dolphin Discovery' but more of that later.

After the meeting, we decided to make our way down to the pool bar where, 'Asterix' (sorry 'Asterio'), the barman was serving drinks to guests who were seated on underwater stools. We managed to find two adjoining free seats and joined in the fun. At this point, it 'struck me' (as a lot of things do because I am accident-prone and because we were sitting at a bar half submerged), that this was a wonderful concept. If urine was blue (instead of yellow), it would be possible to create perpetual drinking with no need to waste time by going to the loo as no one would know you were increasing the pool water level.

There were, of course, other organised activities each day as well as the use of non-motorised sports, and of course, the obligatory, evening entertainment. However, in the case of the latter, I use the term 'entertainment' in a very loose sense, as it was beaten into second place by towel drying. We nevertheless overcame this savage amusement and made friends with a couple who were from England and who, believe it or not, lived only half a mile from us. As a result, we liaised with them quite a lot during our stay.

Our hotel served food and drink all day long and you only had to move location, as the various food stations opened and closed, depending on the time. There was nevertheless a price to pay for all this hedonism and it came courtesy of our Mexican friend 'Montezuma' who delivered us (and quite a few other guests) diar…diarhor…the skitters! It was during one of my lengthy loo visits that I penned the following little ditty which should be sung to the tune of 'Pop goes the weasel'.

Half a kilo of chilli and rice,
Half a kilo of guacamole.
Wash it all down with Mexican beer,
And pop! goes your arse-ole.

Well, you've got to do something when you're suffering on the pan.

It was accordingly a few more days into our holiday before any external soirees were risked but when we did venture out, we took the local bus down to one of the big American-style shopping malls to 'sus' out the merchandise. These buses were the most popular form of local transport, at that time, and appeared to have been built by Lego as everything was plastic including the seats which we kept sticking too because of the high humidity. The mind boggled at the thoughts who may have sat on them before us.

One local feature, when using the buses, was the regular serenading by wandering minstrels who jumped on and off between stops in order to collect whatever tips they could glean from passengers. The tunes they played were about as catchy as infectious diseases. One poor individual did his best to amuse (and beguile) us by singing some Mexican garbage whilst shaking one maraca. It was pitiful and I felt like giving him some money just to shut him up. We decided however to grin and bear it as we only had two more stops to endure before we reached our chosen destination. Judging by the reactions from the other passengers, I would estimate that this individual was set for a very short life in the entertainment business and would probably have to turn to a life of crime to make a living. Mind you with the prices as they were in Cancun, he would need to carry off two successful bank robberies in order to buy himself another maraca (to make up a pair) and improve his current career.

When we did eventually reach one of the shopping plazas it was, thankfully, air-conditioned. This was a blessed relief from the furnace of heat that was raging outside, making even the cactus wilt. What met our gaze inside absolutely astounded us, as the place was so high-tech and designer goods orientated, that it took our breath away. Here we were expecting flea markets and 'Speedy Gonzales' bazaars with, proprietors offering 'leetle souveniers' for the 'lady' at a price geared for haggling down to 30% off the original asking price. Let me assure, you folks, that the reality was more akin to Milan, New York or Paris with designer outlets for Gucci, Eves San Laurent and Versace to name but a few. There was technology everywhere the eye could see, with internet café's, micro-chip shopping emporiums, high-tech gadget stores and very, very expensive jewellery shops.

Gone are the days of buying postcards and posting them to your friends and family back home. The new idiom was to create your own personalised-e card with digital photography superimposed upon computer-generated backgrounds, and then attaching them to emails and sending the finished product anywhere in the world via the internet. You could even add video footage at a slightly increased cost. Please remember, folks, that this was happening over twenty years ago. Even cafes and restaurants were super high-tech with the issue of a micro-chip tag that you handed to the sales staff (at various food stalls). There they simply slotted your chip tag into their machine to register the transaction before handing it back to you. When you had completed your choices and consumed your meal, you simply took your tag to the operator at the main computer exit point where the cashier totalled your final bill. This was the only point at which money or bank cards were used.

You could have your portrait painted by a computer, a personalised calendar prepared, your own personalised business cards (and stationary), toilet paper with your photo or name on it, or even take a trip into the world of virtual reality. You could also have your perfect partner designed by computer and stored on disk (for use at home) should you ever wish to escape to a world of pure dominance without them ever answering back. We were stunned by all this technology and wondered where the hell we were because it certainly didn't seem like Mexico or anywhere else on this planet to be truthful. The whole experience was surreal and put us both off ever coming back.

We had been expecting to be bothered by mariachi bands, time share sales people, beggars, fast mice, cactus plants and people wearing ‘kiss me quick’ sombreros with droopy moustaches. Here, instead, the place was more like ‘In-the-net’ and ‘WWW’ stood for ‘We Will Wipe-out’ your credit balance and leave you with the ‘Dot’.

Chapter 13
Dredger's Rash

The nearest we came to some resemblance of Mexico was the hotel 'Maitre Di' who was called Juanita; strange, I thought, because she actually had two teeth. (Yeah, I know! But I couldn't help myself.) She was always dressed in traditional clothing at every meal time. Then, of course there was the hotel donkey, who I nicknamed 'Otay' (donkey-otay…ou'll have to keep up here because I am on a roll). I thought up the name after that famous Spanish idiot who kept trying to declare war on windmills amongst other things. The poor animal was paraded into the hotel foyer every evening by his three Mexican keepers who were virtually assaulting the guests (with requests for photographs) on their way into the dining room.

As we passed them, I was accosted and had a rifle thrust in my hand, a poncho draped over my shoulders and a sombrero placed on my head. This was before I could say 'Ariba' which of course is what they wanted me to shout as they carried out the photo shoot. Then without so much as a blink they handed me a bill for 200 pesos. Well, I don't have to tell you how mad I got but I will anyway. I puffed out my chest, stared him in the face and told him boldly that our room was number 417. It was actually 217 and I had to grow a beard to disguise myself for the rest of the holiday. This. I may add, was mainly for protection from the occupants of room 417 who were being harassed by one of the donkey's owners.

So much for British gall! Boy, I really showed them.

My Spanish linguistic skills left a lot to be desired and my attempts to make myself known to members of the hotel staff brought me dangerously close to being set on. One such occasion, as an example, was when I tried to obtain two of the yellow-coloured beach towels (from the towel guy near the pool), it was clear that I had said something that did not go down too well with him judging by the look on his face. For all I know, I may have suggested that I had slept with his wife or worse. Incidents like this were starting to become too frequent for my liking so I decided to stick to broken English and finger maths whilst pretending to be stupid. This latter disposition was second nature to me according to Mrs. J.

I did however manage to make friends with the bar staff and even learnt their names off by heart. There was Ishmael and Asterio (at the pool bar) and Carlos (the jackal), Pedro and Wilbert looking after the hotel facilities. Wilbert told me he used to be a fireman but had packed it in when his wife had given birth to twins. I was sorely tempted, as a joke, to ask if they had called them Jose and Hose B (sorry, couldn't resist that) but resisted as I am sure he had heard it loads of times. All the relevant staff soon committed our usual drinks orders to memory and a single upright pointed finger was sufficient for the order of one drink (meaning for me alone) whilst two airborne digits indicated drinks for the two of us. They even knew our preferred measures of alcohol and the right amount of mixer to go with it and this proved to be a wonderful form of communication until our drinks preferences were changed.

I was grateful that the hotel was 'all-inclusive' because you did not have to carry cash around with you. This was perfect because the Mexican Peso is a strange currency with frequent rate changes. It is fine if you are exchanging US dollars but quite the contrary when you are offering British pounds sterling as few places were keen to handle it. This meant that in most cases, when shopping outside the hotel, we had to visit the old town and use the local banks to obtain anywhere near a decent exchange rate. These banks, however, were notoriously slow (even for bank robbers) and great snake like queues developed inside the premises before spilling out onto the pavement outside. These queues were never straight and often coiled around in ever-decreasing circles until they were forced into a dog-leg that led outside. It was sheer chaos and a welcome relief when we were eventually served. Mind you, we still had to negotiate the maze in order to exit the building.

The holiday was not all hassle and we experienced some high spots including my first attempts at snorkelling. Unfortunately, my mask and air-tube were not a good fit and caused me to frequently stand up in order to release the collected water and breathe properly. After a while, I did manage to secure both slightly better and I swam off like Jacques Cousteau in search of sunken treasure, tropical fish and other Caribbean Sea creatures. At one point, I hit an area of shallow water near a sand bank and, due to overeating, my belly scraped the seabed and disturbed the sand. I suddenly felt a sharp pain in the region just above my navel and quickly shot out of the water and onto my hands and knees. The shock had caused me to inhale some seawater and this was rapidly spat out amidst a bout of coughing and wheezing as I gasped for air. My good wife, who had been wading in the sea nearby, burst out laughing as she said I reminded her of the Goon Show character called 'Eccles'. I indicated to her that something had just bitten me and pointed to an increasing circle of redness around my belly button. She did not initially believe me and suggested I was suffering from 'Dredgers Rash'. People can be so cruel at times like this.

It later transpired, after consulting a nearby billboard, located on the beach area and identifying local marine life, that I had been bitten by a strange-looking critter called a 'Cowfish'. It was so named because it had two horn-like stingers on its head. These little buggers lie just below the sand where they are camouflaged from their prey, which in this instance was me. It transpired that their bites are relatively harmless so I eventually plucked up the courage to give the snorkelling another try. I managed to witness many other beautiful species of fish ranging in colour from deep blue to yellow as well as some bright coloured coral, a huge starfish and even a barracuda like fish that I kept well away from. Having already been bitten by a cowfish I was beginning to think that the undersea fraternity were considering me as bait.

Another highlight of the trip was our visit to 'The Dolphin Discovery' that was located on the 'Isla Mujeres' (meaning the Island of Women). The trip was an organised excursion and was the most expensive on offer but, hang the expense I thought, Mrs J is worth it and after all, it was a special birthday gift. We had to take a minibus to the designated point where we could board a special high-speed launch. This boat whisked us over the most intensely blue sea water I have ever seen, towards the island where the experience was located. If I did not know better, I would swear it had been diluted with 'blue curacoa'. Once we had docked on the island we were escorted towards the 'Dolphin Discovery' premises to join loads of other tourists awaiting the experience of a lifetime.

When our appointed time arrived, we were taken down a ramp to some large sea filled pens where a total of eight dolphins were housed. There was also a proviso that you could only swim with them if you had used bio-degradable sun lotion, which we had due to advanced notification. The arrangement at the pens was that only four dolphins would be used at any one time and of these four, two were assigned to a party of six visitors and the other two to another party of six. In our group, the dolphin trainer was a girl called Andrea, who I discovered was from Manchester in England and who had (as she stated) got the job by being in the right place at the right time. Our two dolphins were called Danielle and Trojan and we were asked to kiss each one in turn, on its snout, and in return they would kiss us back.

I had no problem with Danielle, as she was quite attractive, but refused to be photographed sticking the lips on Trojan as I feared the negatives. Such a shot could be used on the internet and totally damage my street cred which was already starting to diminish with my previous escapades. Despite this, we entered into the spirit of things and started to carry out the various requests the handler asked of us. Before long, we were being lifted out of the water by the two dolphins who used their snouts to push up our feet and carry us forward. This was a very surreal experience. We were also pulled along by their dorsal fins and held hoops in the air for them to jump through. The whole event was video-taped and we were given the option to purchase a copy for our own cherished memories. This whole trip had surpassed our wildest expectations and went down as one of life's high spots which we would recommend to anyone.

Although I was now clear of the Montezuma problem, I had nevertheless developed a form of temporary deafness which I believed to have been an after effect of my snorkelling adventures. This I am happy to report was something that I could easily cope with and actually found the condition extremely useful when we participated in the hotels excuse for evening entertainment. It felt like I was in a world of my own as everyone was talking as if they were speaking through paper covered combs. The remainder of our holiday was spent topping up our suntan and were by this time a golden brown and down to using factor four lotion albeit that I was applying it quite thickly and more often than before. Mrs J was annoying me by nagging on about the quantities I was using and even asked at one point, if I was contemplating a cross-channel swim. This, folks, is the sort of humour I have to endure from her but it goes with the territory.

My hedonistic tendencies continued with such luxuries as ordering double (or even triple) measures of spirits and consuming snacks such as chicken wings and pork ribs whilst lying in the hotel jacuzzi. Well, this is what holidaying is all about if you ask me. I mean who wants to visit exotic locations to undertake studies of sewerage problems or educational trends in rural villages to name but a few.

And so it came to pass that we successfully reached the end of our Mexican vacation without being discovered by the donkey owner or the occupants of room 417 and found ourselves ensconced in the departure lounge of the airport ready for our long flight back home. Alas, our flight was delayed and we were forced to spend several hours in the terminal building trying to amuse ourselves. This took the form of crossword puzzles, duty-free shopping (without buying anything), gift shop scanning and any other exciting ways we could think of, to make the time go quicker. I was getting really bored and even resorted to going into the gent's toilets which was so big that I decided to count the urinals. I can report that in fact there was a total of 24 arranged in three sections of eight, the final section of which was in an annexe running at right angles. I was one of the biggest airport loos I had ever seen and could not wait to pass on the result of my research findings to Iris and inform her of the discovery. Judging by her comments, it seemed she was ready to have me sectioned under The Mental Health Act. There is just no pleasing some people.

When the message to proceed to the boarding gate was eventually called, we were more than relieved as tiredness was beginning to set in. After we had boarded the plane, we were once again sprayed with a fine mist. The captain spoke to all the passengers over the plane's tannoy system and tried to alleviate concerns by stating that there was nothing to worry about as the mist was only the effects of cabin pressurisation due to the mix of humid air meeting cold air from the aircraft vents. No one, including us, believed this and I complained to Mrs J that if we ended up deformed in later life, by growing another head or something similar, then I was going to sue the airline. In fact, both my heads would.

Chapter 14
Interlude

I am calling 'time out' at this juncture for two reasons. The first is because I want to point out that we have taken too many holidays to include in this book, and to do so may prove counter-productive. The second is that I wish to go off-piste and add one particular break that we took with our two kids purely because it is worthy of a mention. You will find this in the following chapters 15 and 16.

Of those aforementioned holidays, we have been to many parts worldwide such as Australia, New Zealand, Thailand, Hong Kong and China, Norway, Finland, Sweden, Denmark, Russia and the Baltic states, most European destinations, Africa, Japan and the USA, to name but a few. We have been lucky enough on these travels to visit some spectacular sights such as The Grand Canyon, The Great Wall of China and The Pyramids at Giza, Niagara Falls and Petra in Jordan, to name but a few.

Added to the above list, we can also include several cruises that have traversed the main oceans where such exotic locales as the Hawaii Islands, Panama Canal, The Azores, South Pacific Islands and (as mentioned previously) the east coast of South America were visited along with many others not so exotic. Sadly, these cruises have been taken during our ageing years and for some reason the sparks of humorous encounters have been limited. I am not certain as to the roots of this but it may be due, in the main, to a much older clientele that have perhaps behaved themselves slightly more when in our presence. This seems to be a pattern of most cruise holidays.

Although these locations are both exciting (and exotic), they did not always provide sufficient humour for the purposes of the book. Also, there is the fact that I did not take copious notes of our cruise encounters and as my memory is not what it used to be, I would probably have to resort to telling more than a few fibs and that would be a tad dishonest.

Nevertheless, there are a few interesting moments that could be spoken of and I thought that perhaps I could add these as a relevant conclusion of our previous travel history. In this regard, I am including them in a few later chapters under the headings of 'Relevant Snippets' and 'Worthy Mentions' which you can locate at Chapter 17 onwards.

Chapter 15
The Itchy – Richter Scale

I have always had a long-held passions for boats and all things nautical so it is of no surprise that sooner or later we would end up taking a boating holiday. To this end, we decided to book a family holiday on one of England's finest inland waterways, namely 'The Norfolk Broads'. With our daughter Caroline (who would have been aged 16 years at the time) and our son Andrew (who was 10 years old), we set off in true 'Swallows an Amazons' style to take collection of our pre-booked floating home for the following seven days.

Our journey from the north-east to the Norfolk area (at that time in 1996) was long and tiresome, with the last leg of the trip especially so, as the roads seemed to get narrower and the traffic heavier as we neared the area. I assume they have cottoned on by now with the introduction of bypasses or some other means of speeding up the travel problems. Despite these setbacks we eventually arrived, on time, at our allotted boatyard which was located at a place called 'Acle Bridge' ('Acle' being pronounced 'Eye-Kull' by the locals). We managed to find one of the staff who checked our documentation and showed us where to park our car for the duration of the holiday. We collected our luggage and made our way along a timber jetty where a magnificent 'thirty-six-foot' floating home was waiting to welcome us on board.

The problem was that it was the type of craft that you steer from the front and, as I was to learn later on, this becomes a severe handicap when reversing the boat, particularly against the current, but that saga will have to wait. Before we were allowed to leave the boatyard, we were given brief instructions on how to start up the engine (as well as how to stop it), steering protocols and how to empty the toilet waste. The only other information imparted to us related to the filling of the fuel tank and how to moor up during and at the end of the day. I have to admit that very little of this info was being absorbed into my brain due to the excitement of casting off and getting afloat. This, as it turned out, was my first big mistake.

When our nautical adventures eventually got underway, we had no idea where we were going and (even if we did) we had no idea of how to get there. The only map we possessed was a tatty affair and was so old that it showed Viking landings and Boudica's last known encampment. I may be exaggerating slightly but you get my drift. Undaunted, we set off in the direction that the boat was pointing which, as it turned out, was to take us towards Ye Olde Village of Wroxham where Sir Roy of Norfolk had his land and all chattels upon it.

I have no idea what Wroxham looks like today but back then every business was named Roy's. There was Roy's Garage, Roy's Convenience Store, Roy's Butchers, Roy's Greengrocer's and so on. When we arrived there, we got to learn that if you were called 'Roy' you could own your own shops, garages, warehouses, jetty's, windmills and just about anything that generated income. Anyone who visited the place years ago will know that Roy was God in Wroxham and everyone else was merely Joe Public. We learnt that Roy is extremely elusive and there are doubts that he exists at all. We never saw him in person and any questions you asked about him were met with stony silence from the locals. Having said that it could have been our north-east accents or the fact that down that way he was probably referred to as Mr. President.

Anyway, I digress and need to get back to the boating. Once we had set off, it was a wonderful experience being in charge of a floating home that was of sufficient length and weight that it had the capabilities to demolish other craft, waterside dwellings and even the odd jetty or two. One of the things in our favour was that the Norfolk Broads has an area of around 125 miles of lock free navigable waterways. This however has to be tempered with the fact that there are loads of other boats using the same routes and even the possibility of submerged tree roots and branches should you stray from the main channels. You also had to stick to the right side of the channels unless you were overtaking even when sticking to the very slow speed limits.

For the uninitiated, the Norfolk Broads are not 'Fen-land whores' but a man-made phenomenon caused by its medieval inhabitants digging up vast amounts of peat to use for their fires. Following these considerable excavations, the whole area flooded when sea levels rose during the fourteenth century, and this resulted in 40 or more shallow lakes (broads) being formed, making this such a unique location. It just so happens that Wroxham was saved from being flooded probably because Sir Roy owned the peat works and hired the men to forage for fuel well away from his place of abode. I reckon Roy was really the English version of Noah and he had some divine knowledge about the 'up and coming' sea level changes, at that time.

I hope you are well impressed by this local knowledge because there is more. Because of the area's importance to wildlife, it gained national park status in 1989 and includes five rivers namely 'The Bure', 'The Yare', 'The Ant', 'The Thurne' and 'The Waveney'. There is also 'The River Chet' but is actually a tributary of 'Yare'. Traversing the rivers and broads requires the user to abide by a set of simple rules. The speed limit was 5mph (now believed to be even less in some places) but there were only a few who actually stuck to this and I must admit that I found it very difficult to maintain such a slow pace and exceeded this on quite a few occasions in order to reach our chosen destination before nightfall. The other main points are that all motorised river craft must port to port (left to left) when passing head on and that any sailing vessels had a right of way over powered craft. It was also necessary to know the correct use of the boats horn. This I can break down as follows.

1 short blast = moving to starboard (right), 2 short blasts = moving to port (left), 3 short blasts = reversing, 4 short blasts = unable to manoeuvre, 1 long blast is used when going under bridges, and 1 attempted signal without a blast means the horn is broken.

Armed with all this useful information, we finally set course along the River Bure travelling in a northerly direction which took us through some pretty boring scenery. The pace was fairly leisurely and I was getting quite blasé about handling the boat, so much so that I had resorted to one handed steering whilst drinking a can of pop. This honeymoon phase ended suddenly when we approached the junction with the mouth of 'The River Thurne'. Here there were boats of all sizes and descriptions entering into the 'River Bure' passing and heading both ways along it. To us inexperienced sailors, it felt like our narrow country lane had just become a main motorway. It was quite a nerve-racking experience for an amateur boating party fresh from the boatyard.

We nevertheless accomplished the through passage and carried our journey along the Bure until we reached the turning for the 'River Ant'. After negotiating this manoeuvre, we came across the next big challenge namely 'Ludham Bridge' where we were held up awaiting our turn to enter and exit onwards to the small village of 'Irstead'. After a while, things once again became a little more relaxed and the area seemed more picturesque. This made up for the earlier hassles and made us feel considerably more confident and thus take in the riverside panorama at our own pace.

The next main feature was in fact the main purpose of our journey up the 'River Ant', namely Barton Broad (our first Broad). Here all boats had to keep strictly to marked channels that incidentally encompassed a small island named 'Pleasure Hill.' It was here that Mrs J wanted me to pull alongside and stop for sufficient time to allow us to send a landing party ashore for some general pleasure! I however had other plans and as captain I over-ruled the crew who I must say were becoming a little restless at the time, but not quite at the point of mutiny. I felt that it was more important that we find ourselves a suitable mooring for the night before the majority of the decent ones were snapped up. Although unpopular, someone has to make these decisions and that is what rank is for. Sadly, we found that all had already been taken.

Undaunted we set off back along the ‘River Ant’ until we reached the junction with the Bure. We then travelled along towards Malthouse Broad to see if we could find a suitable mooring there. This small broad (which adjoins ‘Ranworth Broad’) was ideal for our purposes and we luckily found a small inlet (or Dyke) for our first overnight stay. I was glad of this respite because I was becoming extremely tired from the days travelling (by car) and the subsequent responsibilities of command. Whilst Mrs J set to rustling us up some dinner I entertained the kids with a spot of fishing. As this sideline failed to produce anything edible to go with chips, we had to settle for sausage and eggs as the accompaniment.

I have to say however that this feast married extremely with a cheeky little chardonnay from the Chilean foothills, albeit that it was served lukewarm. This wine carried a subtle hint of something that we were unable to determine (with our limited palettes) but I am sure that Jilly Goolden and Oz Clarke would be spouting from a thesaurus. Oh! The delights of England’s dirty green and pleasant waterways.

Our first night afloat passed with a sleep broken (now and again) with a cacophony of strange sounds emanating from somewhere nearby our boat and may not have been natural. It was, after all, very difficult to distinguish between the flushing of a nearby chemical toilet, the surfacing of marsh gas bubbles or the mating calls of the tufted duck. With the dawn came a whole chorus of new sounds that included bird song, human voices, boat engines and the lapping of small waves against the side of our vessel. Yes, folks! We had slept in and almost everyone else had set off to explore pastures new (or should I say watery pastures). Undaunted we emerged from our slumbers full of excitement and very hungry. This latter condition was soon resolved by a hearty breakfast of bacon and eggs washed down with copious amounts of hot tea. Lovely jubbly.

When the time came, we cast off and resumed our journey back along the 'River Bure' and onto Wroxham, the unofficial capital of the Norfolk Broads. Along the way, we passed through the beautiful setting of Horning with its numerous riverside homes, gardens, hotels, cafes and shops. Unfortunately, Horning was too busy a place to stop, and in any event finding a mooring area was extremely difficult. And so, we decided to carry on past 'Dydler's Mill' (the mind boggles what went on there) and onto Wroxham Broad where there was plenty of places to moor up and get off for a walk. The walk we took led us through some delightful country lanes and into the outskirts of the town of Wroxham. We did not go all the way as the day was too nice to waste by looking in shop windows and mingling with crowds of other holiday makers, day trippers and locals. Instead, we turned back and discovered lots of lovely lanes to walk along and relax. After a while, we returned to our boat for some lunch and then later decided to try a spot of swimming in the shallow area of the broad. The kids loved it but Iris and I considered the water to be a little too cold for our liking. Eventually, we all returned to our boat and settled down for a spot of reading and relaxation.

The rest of the day passed peacefully and after tea, we decided to explore the rest of the broad, which was in fact quite substantial in size. By the time we returned to our original mooring, there was a whole flotilla of other vessels all moored in a fashion that would not have looked out of place at the 'Cote de Zur'. Only one vacant slot remained and this meant that, if we wanted it, we would have to reverse into what seemed like an impossibly small mooring. Undaunted I decided to take on the challenge and positioned the boat for rear entry. I dispensed with the 3 short blasts on the horn and slowly reversed our boat whilst relying on the crew to guide me in.

This was mistake number two. I was blind to the narrow opening between two other craft and it was only a muted cry from astern that alerted me to the fact that a collision was imminent. I quickly engaged forward gear and revved the engine hard. This saved us and a second attempt was undertaken that sadly was no better than the first. After two more attempts, I decided to pull rank and rather embarrassingly headed off towards a small island adjoining the thoroughfare of the 'River Bure'. It was here a frontal approach could be made for a mooring and in doing so I had to ignore the cheers, and applause, of the cretins who had moored successfully. I convinced myself (in the process) that I would have been capable of doing it first time had I the benefit of rear cockpit steering as they all did. I would get my own back. I just needed time to work out how to do it.

Once moored (in our new locale), we discovered the reason why no one else had used this space and that was that the island was overcrowded with Canada Geese who are a noisy bunch to say the least. They additionally hold the local records for shitting green poo everywhere they waddle, even on boats that foolishly moor up to their island. Our other nagging problem was of course the midges and other irritating insects that are attracted by the damp conditions, and, let's be fair places do not come much damper than the Norfolk Broads. Before we had set off on this holiday, I had taken the trouble to visit my local GP to ask for advice about avoiding diseases from biting insects. He told me it was best not to bite any of them. Oh! I do hate doctors with a sense of humour, don't you? Seriously though I had obtained all the necessary creams and sprays from the chemists and felt suitably safeguarded against any of 'Mother Nature's' pests.

However, it should be noted that the Norfolk Midge is second only to the Scottish variety, in terms of annoyance, and falls somewhere between seven and eight on the 'Itchy-Richter' scale. They have a profound strategy that is clearly governed by an intelligence base on Decoy Broad. Swarms of the little buggers brought back memories of their Canadian cousins and, just like them, were spoiling our newfound boating experience. I started to wonder what else this peculiar environment was going to throw at us. Perhaps 'willow rash', 'Fen-land croup' or even 'rush cutters chin'.

Chapter 16
If at First, You Don't Succeed...Try Doing It the Way Your Wife Tells You

Despite suffering some general discomforts, we were now well and truly into our holiday and actually starting to enjoy the whole boating experience. No longer were we the 'new kids on the block' but seasoned mariners capable of negotiating low bridge arches, narrow waterways, tidal flows, local fisher folk and mooring up for the night. I left this latter skill until the end because it did, to say the least, cause us some embarrassing moments and at one point the marooning of my wife and daughter who were tasked with affixing the grappling hooks to a soft embankment. You should know (at this point) that according to Mrs J, I have only got two faults—'Everything I say' and 'Everything I do'. As such she was less than pleased when she had to leap from our moving river cruiser that was reversing (against the tidal flow and a very strong wind). What happened is difficult to explain but I will give you the best account that my feeble mind can recall.

Iris was the first to set foot (only one) on the embankment and made ready to grab our daughter Caroline when she jumped off the boat onto the grass. Now, at this point, I had foolishly come to the conclusion that, whatever happened, the laws of gravity would assist Mrs J to maintain a balance on the bank once Caroline had reached ‘terra firma’. All however did not go to plan (probably because that law was not recognised by Sir Roy of Wroxham and therefore did not exist in these parts) and Mrs J ended up with one foot on the bank and one foot on the boat. I conjured up visions of her doing the splits before ending up drowned, and then me having to bury her in a ‘Y’-shaped coffin. However, after a few minutes of high-pitched screaming (mostly at me), I managed to manoeuvre the boat to a point where she was able to jump clear and end up with both feet on the bank.

Then, of course, was the problem of hanging onto the mooring ropes which by this time had become fairly taught and were likely to pull both of the landing party into the water. After a few minutes, Iris grew a little exasperated (for the uninitiated that is a small flower related to the delphinium) and called me a f…ing idiot and gestured to me with two fingers which I took to mean that both grappling hooks were in place. You know, it’s times like these that a strong marital understanding comes in useful and this proved to be the start of a greater link between captain and crew. This understanding was to be the basis that would save us from impending disaster on more than one occasion.

Boating on inland waterways is such a relaxed way of life that it is easy to forget that the floating object under your control is capable of doing untold damage and causing harrowing experiences for other boat users not prepared for the Geordie Navy. The only thing I lacked as captain was the appropriate headgear and had to make do with a knotted handkerchief as a symbol of my status. We had also brought with us a small pennant, that we erected at the front of our craft, and this signalled to others that we were a special boating party and needed to be given priority. The pennant proudly displayed the skull and crossbones, our family crest.

Yes, we were a proud bunch and I had the added distinction of being from noble stock. Indeed, my grandfather (on my father's side), was nearly a 'peer' and coincidentally my grandmother had also suffered with kidney trouble. Only joking! With a background like that, it was only right that I should take command of the vessel for the week and allow the rest of the family to act as crew. Alas, however, the pressure of command was taking its toll and I had to endure much abuse from 'her on deck' in order to maintain an equilibrium. As they say, *If at first you don't succeed, try doing it the way your wife tells you.*

And so, it came to pass (as did many other boats), that the crew became restless for contact with other human beings and I was pressurised into mooring the boat at Wroxham in order to facilitate a landing and the chance to shop for supplies and take in the sights of a new settlement. It occurred to me that Roy was in need of a visit from the 'Monopolies Commission' as he seemed to own just about every damn business in this place. This saturation of his commercial enterprise had become known (locally) as 'the largest village store in the world' and goes to prove what value there is in having a large family of business orientated mafia. At the risk of repeating myself, it is worth noting that here was Roy's Bakery, Roy's greengrocers, Roy's garage, Roy's DIY, Roy's garden centre, Roy's department store, Miss Roy (the mind boggles), Roy's coffee shop, Roy's taxis and Roy's supermarket. This was all

I can remember and perhaps there is even Roy's car park, Roy's undertakers and maybe even Roy's brothel. It was surprising that boat building, travel agents, local funfair, wedding shop and rent a car did not feature as a 'cradle to grave' policy that could easily have been adopted to add to the overall retail ambience. Anyway, I'm sure you get the drift.

We did venture into one of his premises to purchase a newspaper and some matches. These incidentally were a copy of *The Daily Roy* and a box of Roy Vestas. Seriously, we did manage to collect supplies of food and other essentials before returning to our boat and heading back along the River Bure towards South Walsham Broad for another overnight stay. Along the route, we spotted lots of our feathered friends that we were able to identify from a book of British Birds loaned from our local library back home. You know, I think that I was born to be a 'twitcher' but never really got into the hobby because of the effects it may have on my street cred. Nevertheless, we were able to spot a kingfisher, a grey heron, some mute swans, several species of duck, or waterfowl, and even some 'tits' which nearly caused me to crash the boat. I have always liked tits and they are definitely my favourite.

That particular night passed uneventfully and we were able to gain a good sleep, probably our first. The next morning, we set off after an early breakfast and once in the main channel, I decided to relinquish command and let Mrs J take the tiller, so to speak. She did really well which allowed me to relax and take in the sights of Broadland as well as a spot of sunbathing which was most welcome. It was not long before we reached the turning for the 'River Thurne' and travelled along towards that other tourist attraction known as

Potter Higham and its infamous bridge. As we approached the bridge, I resumed control of the boat as it was high water and the already low span of the bridge was even lower. I doubt, however, that I would have attempted the through passage even if they had drained the river and offered us logs to roll through on.

Perhaps it was the number of notches on the arch of the bridge or the numerous small dashes of boat paint and varnish that put me off. Whatever the reason, I decided that trying to squeeze 36 foot of fibreglass through an archway seven feet high, at its best, was not for me. And so, with a triple toot on the horn, and a quick reverse action leading to a smooth three-point turn, we headed back along the 'River Thurne' and settled for a detour up along Womack Water. This to our dismay was not worth the visit and so reluctantly we turned back towards the 'River Bure' and headed for an overnight mooring near Acle Bridge.

This section of the broads is fairly busy and seems to act as a main thoroughfare for most types of craft. Whilst we were there, we had the pleasure of observing one of the famous Norfolk Wherry's. Now I feel that at this point I should tell you a little about the wherry because it has a long history. It was primarily a trading craft that evolved to suit the conditions encountered on the waterways of this part of the world. Its predecessor was 'The Keel', a square-rigged sailing barge dating back to the time of 'William the Conker' (who incidentally was a friend of Sir Roy). The difference however with the Norfolk Wherry is that it differs in that it sports a high-peaked gaff, with the mast well forward and of special design. Unstayed, it pivots at the top of the tabernacle and was counterbalanced by an enormous metal weight of up to two tons. This meant that it could be lowered quickly in order to pass under bridges, then raised again without the craft losing way. I hope you are suitably impressed by all this because I now have a headache.

These wherries were crewed by a skipper and mate (who apparently may also have been his wife—or 'mistress') nudge, nudge, say no more, wink, wink. Sadly, by 1949, the last working 'wherry' had finally succumbed to road and rail transport and accordingly furled its sails for the final time and ceased trading. Most of these crafts were scrapped before their worth was realised except for one 'The Albion' and this was the one we had the pleasure of witnessing albeit as a sailing experience for enthusiasts and other visitors. It's surprising that P&O don't adopt them as cross-channel wherries!

Anyway, let's get back to the story. The rest of that day passed peacefully although I did have a few anxious moments worrying about whether I had made the appropriate allowances with the mooring ropes to be able to cope with the tidal variances on this stretch of river. Another unfortunate thing about this stretch of water is that it is boring as it lacked any natural features. To make matters worse, it had started to rain and it felt decidedly colder. Mrs J started grumbling, as she usually does whenever the temperature drops, and the kids were getting bored with sitting around. Fishing was out of the question, as was walking and cycling (the latter pursuit was dogged by the fact that we did not have any bikes—a small technicality). This only left a game of cards and even that was confined to 'snap'.

Boredom eventually led to sleep after darkness had descended and before you could say, "Bob's got scurvy," it was daylight again and the delights of breakfast were once again upon us. Even the weather had improved and all was looking rosy once again. Because of this, we decided to venture down the Bure to Great Yarmouth and make it our final port of call for this holiday. Time was alas running out and we calculated that we had just enough time left to make a brief visit and return to our starting location at Acle Bridge. We checked the route carefully and set off into what was probably the most hazardous stretch of the broads in terms of tidal water, concomitant currents, heavy boat traffic and the dangers of varying water depths.

It is on this stretch of water that you pass by the charming village of 'Stokesby' with its redbrick riverside cottages and the thatched church of St Andrew. I've heard that it has some excellent brasses for those who are into rubbing, and to top it all there is a small workshop that specialises in candle making. What more could you want? It was very tempting to moor up at 'Stokesby' but alas, there was a severe lack of free space, so sadly we never managed a stop. Obviously, everyone who was passing had the same idea and luckily for some of them, they arrived earlier and grabbed all the free spots. So, it was onwards and upwards (river that is) until we heard the raucous shrieks of seabirds overhead. At this point, we realised that the port of Great Yarmouth was close by and I gave the cry, "All hands, on deck!" with a lookout posted at the bow.

Even the air had taken on a salty tang and the ozone was considerably fresher and it was not long before the area became more built up with riverside buildings most of which were commercial in style and industrious in parts. The 'River Bure' was much harsher and less forgiving than at Wroxham and Horning and waves were apparent due to the greater tidal flow that was present on this particular section. In addition, the number of river crafts had increased and water traffic was quite heavy (in both directions) as a result. The river was also getting decidedly narrower at this point and greater skill was needed to navigate a safe passage. We were not in any position to turn around, so it became inevitable that we had to pass under two road bridges before reaching any stretch of water wide enough to consider a turnaround. The next phase was to turn to starboard (right) at a yellow-coloured dolphin marker which took us onto the 'River Yare' and heading into a wide expanse known as 'Breydon Water' which almost felt like a lake compared to what we had just negotiated.

This expanse of water is however quite shallow and not suitable for all types of craft. It was therefore essential that you steered between the red and green coloured markers that indicated the deeper water channels. We decided to travel as far as the 'Berney Arms' pub and avoid the 'River Waveney' because we had no time left of our holiday in which to explore it. We managed to find a mooring spot close by and availed ourselves of the pubs hospitality as it is claimed to be the most remote Broadland pub and was, according to our guidebook, only accessible by river or railway.

Near this pub is the distinctive landmark of 'Berney Arms Mill' which is a fully restored drainage mill that can be explored by visitors. We did however give this attraction a miss and decided to return to our vessel to partake in a spot of fishing. Incidentally, we still did not catch anything remotely aquatic and I put it down to the fact that the fish around these parts are spoilt and did not like our mealtime leftovers.

And so it came to pass that in the morning, we departed for the Acle Bridge boatyard and finally put an end to our great floating adventure. As we travelled back to base, I felt that I had become a seasoned river captain and had gained a great deal of navigational skills to boot. My suggestion that we should save up and buy our own boat was met with one of those harsh stares from Mrs J that roughly translated into a resounding 'P…s off', no matter what language you spoke.

When we did eventually arrive at the boatyard, we conducted the usual handing over ceremony before bidding them farewell and setting off for home in the car. It was agreed by all that the trip had indeed been different and in some parts enjoyable but alas, no one (but me) seemed to want to repeat it, or indeed, visit this neck of the woods for a long time.

Chapter 17
Snippets and Other Worthy Mentions of Our Travels

Following the 'flying from the nest' (departure) of our kids (for various reasons), and with the onslaught of age and retirement, we made it our intention to continue our marital travels to as many worldwide destinations that took our fancy at the respective and various times in our lives. There are far too many to mention and in the main, they were uneventful and contain less humour as we had mellowed with age and accordingly behaved ourselves a little better. For this reason, the bulk of the relevant journeys are not worthy of mention and somehow do not fit with the ambit of this book.

Of course, there is the odd exception and what follows are some snippets and worthy mentions that may raise the odd titter or two. Instead of full chapters, these inclusions will take the form of short stories from selected expeditions in the hope that these will meet with the criteria of this book. Some are based on air travel holidays whilst others are directly related to cruise packages, or a mix of both.

These snippets somehow occurred without prior purpose or design and any included humour is purely down to naturally occurring events. So, if you have not yet reached that point where you wish to take this book for recycling, to your local charity shop, then please read on and enjoy.

Chapter 18
Snippet 1 Oh! I Do Like to Be Beside the Seaside

Shiver me timbers, we booked another cruise! (This was probably our 11th). Only this time, it was a journey that would take us across the equator in the Atlantic Ocean bound for the shores of Brazil (on the east coast of South America), and in particular to Rio De Janeiro. In order to facilitate this journey, we first had to fly to Ravenna in Italy before transferring to a coach for the road trip to the 'Porto Corsini' cruise terminal. Here we boarded the cruise ship 'Costa Concordia'. Yes! You've guessed it. It was the one that sank on 13 January 2012. Our particular cruise, however, was taken approximately 18 months earlier.

I am not going to bore you with full details of the cruise but instead pick out some selected moments worthy of mention. Now for those of you who have never cruised with Costa, you need to be aware that the passengers are usually multi-national and of course speak several different languages. This leads to some interesting situations such as quizzes and live entertainment. The quizzes in particular took ages as each question had to be read out in seven different languages and after each session (of say 20 questions), the answers had to be read out in a similar fashion. Such sessions accordingly lasted well over an hour by which time, most of us had lost the will to live. In respect of evening entertainment, it fared little better as there were no comedians (as it would be pointless), so it was more than likely a dance troupe or some acrobatic duo, with both type of acts equally boring.

Nevertheless, we cruised the Atlantic and crossed the equator. At this point, the usual ceremony took place with staff dressed up appropriately as King Neptune and Co. and much ducking and diving in the on-board swimming pool was to be had. If you ever get the chance to witness this tradition, it is well worth it. However, our sights were firmly set on Rio and the many delights and tourist attractions that that location is famous for. There were obviously several ports of call before we reached this destination but none carried the international buzz such as that destination did. Not for us anyway.

When the great moment arrived, we were totally dismayed to find we had docked in some extremely poor weather. It was raining heavily and everywhere appeared to be dull and uninviting. There was low cloud everywhere and visibility was impaired to say the least. Nevertheless, the excursions that we and other passengers had booked (in advance) still went ahead regardless and, in order to facilitate this, each of us were issued with knee-length disposable blue plastic poncho's that also had a built-in hood. I am struggling to come up with an accurate description at this point but can only describe them as extremely large condoms.

So it came to pass that our particular party (of 30 blue condoms) set forth for Corcovado Mountain in the Tijuca National Park to access the railway system that would take us up the steep incline that led to the famous statue of 'Christ The Redeemer'. The mountain rail system uses three electrically powered trains and can take up to 540 passengers every hour. The journey by this rail system is quite enjoyable as it passes through some dense jungle like foliage on its route to the summit. Once there, you are directed towards the viewing platform up several steps that lead to the base of the famous statue. Having said all that, the visit was a complete washout (excuse the pun) as all we could see were Christ's feet as the rest of the statue was surrounded by low-lying cloud and was not visible. Similarly, the famous view of the bay was also blanked off so the whole trip was a bit of a disaster.

And so, the 30 blue condoms returned to the base of the mountain for our onward journey to that other famous attraction namely 'Sugarloaf Mountain' where there was a cable car (the one featured in the James Bond movie 'Moonraker') to take us to the summit. Thankfully, the rain had eased off and so had the low-lying cloud, so this element of the excursion was much more agreeable and accordingly we could ditch the condom outfits. I accordingly tried to cheer up Mrs J by stating that the final leg of our excursion was coming up and that was a visit to the famous 'Copacabana' beach.

I remember telling her she would be seeing lots of super fit and tanned men in their skimpy cossies, and that I would equally be casting an eye out for some semi-clad women in bikinis, that the resort had become famous for. However, word must have got out that 30 blue condoms were on their way because when we arrived at the beach, it was deserted. Undaunted we alighted our coach and took a stroll down to the famous zone. I reckon that as I stood on that damp sandy beach, totally alone, I can say hand on heart that it's their loss, not mine.

Chapter 19

Snippet 2 Oh Phuket! Let's go to Thailand and Bangkok

Thailand had been on our radar for some time but the reason we had not been was that it did not appeal as a stand-alone destination. Instead, we decided to include it as a stopover in a once in a lifetime round the world trip lasting just short of five weeks. However, our original plan, to visit Phuket was scuppered as it would have entailed a bit of a dog-leg trip from Bangkok, which to us, had greater appeal at that time of our lives. Anyway, the plan was that we would fly from Newcastle to Heathrow (London) and take a direct flight to Bangkok, where we would have a four day stopover prior to the next leg of the journey which happened to be Perth in Australia, and then beyond.

When we eventually arrived at Bangkok Airport, we took a taxi to our pre-booked hotel on Sukhumvit Road in the heart of Bangkok. This area is known to be a hub for tourists and boasts some of the best shopping, hotels and restaurants in the city. On our first night there, we had dinner in the hotel's rooftop tropical garden restaurant and it was a sublime experience, as we could see the lights of the nearby skyscraper buildings set against a backdrop of a starlit sky. Whilst we tucked into our more than excellent meal, we started thinking that if this was anything to go by, we were set for one ‘hell of a holiday’ experience.

Whilst I do not intend to go through all of our itinerary, there was one memorable event that sticks out in our memories. This occurred on our second full day in the city and commenced early in the day when we decided to take the overhead ‘Skytrain’ rail transport system from Sukhumvit station down towards the river. To our great surprise, the station was practically deserted and when the ‘Skytrain’ finally arrived, with its blacked-out glass windows, we assumed we would be able to find a seat or at least plenty of standing room. Oh! How wrong we were. When the doors opened, it was wall to wall Thai people wedged in like sardines and after some considerable manoeuvring, we too managed to board the train and join the throng. Now I need to mention at this point that I am 6'2" in my socks and Mrs J is 5'10". When we did manage to squeeze into a small space, we felt like we had boarded a train for Lilliput as we were ‘Chest high in Thai’s’. (Blimey! That sounds like a good title for a song or a movie).

What a relief to reach our destination ‘Hua Lamphong’ station (near to the river), which just so happened to be the last stop, as trying to get off at an earlier station would have proved extremely difficult. Anyway, whilst we were at the river, we decided to take a boat trip that was one of the many available to tourists and it proved a welcome respite from the heat of the day. Once that little excursion had ended, we set off on foot towards the area where the temples and palaces were located and it was here that a unique experience commenced.

Whilst we were walking around and taking in the sights, we were approached by a well-dressed gentleman who asked us if we were from the UK. Although he looked to be Thai in origin, he spoke with what can only be described as English (public school) accent. We naturally confirmed that we were and he proceeded to enquire as to whether we would be looking to participate in a city sightseeing tour. Our initial reactions were somewhat negative until he told us that it would only cost 20 Baht which was roughly the equivalent of £2 sterling and that was the price for both of us. As a result, he now had our undivided attention even though we thought there must be a catch.

The catch was that we would be driven around in a tuk-tuk and that all that was required was that we had to visit three or four retail outlets and that way the driver would be able to claim his petrol coupons plus any tips we felt like giving him. We sensed it would probably be hard sell but 'hey-hoh'! We had survived Madeira, so no problem. With our agreement, he raised his hand (like the messiah) and whoosh! A three-wheel 'tuk-tuk' suddenly appeared as if out of thin air. Now this particular tuk-tuk was not new. In fact, I would guess it was several years old and fashioned out of scrap from a time when Thailand was known as Siam. It had three wheels, a canvas roof and a rear seat that just about catered for two people. I reckon it would have been great for those of a vertically challenged nature but for two tall English tourists, it was a tad tight and caused us to have to bend forward and duck our heads.

Without notice, he revved up to a resounding tuk-a-tuk-a tuk and 'Whoosh' away we flew, off into heavy traffic. I had to hang onto the roof of the vehicle and I put my other free arm around Mrs J. She assumed I cared for her safety but in reality, I was using her body to steady myself, but did not let on. The first part of our (contorted) journey was almost a fly past of the palaces and temples that were prevalent in this area. Once this was over, and again without prior warning, our driver headed off into the mainstream traffic towards one of the 'designated shop' stops to collect the first of his precious petrol coupons.

At one point into this journey, he had to stop the tuk-tuk, in the middle of the highway, due to really heavy traffic. Then without so much as a 'Bob's your lung' (Thai for 'Uncle'), he zoomed out into the oncoming traffic and then nipped back in shortly before a head on collision. My knuckles turned white with gripping the roof so hard; although to be fair, they also matched my face as the colour had drained out of that too. At this point, it became clear that petrol coupons were more valuable than life itself and the fee of 20 Baht seemed exorbitant by comparison.

This kamikaze driving was repeated several more times until we pulled into a side street and were invited to alight the vehicle and head into a jewellery shop that seemed to specialise in emeralds. We were shown several special offers on necklaces and bracelets as well as having to endure some high-pressure sales patter. This we managed to do and after they gave up on us, we resumed our cramped journey in the tuk-tuk. Our next stop was to a gent's tailors where I was offered an 'Armani' suit (copy) at a very low price that could be ready in 24 hours. They took my measurements and again subjected me to high-pressure sales pattern which I am delighted to say I sailed through unscathed as we were seasoned veterans.

The next port of call was to a run-down house in a back lane where the driver lived with his family. Here he gestured to us to wait until he popped in to have his lunch. *Cheeky bugger*, we thought especially as we were paying him 20 Baht for the trip. Anyway, we had to endure two more shop stops, I think one was a carpet store but I cannot recall what the other one sold. We did however survive these as well and once the driver had secured his final petrol coupon, he dropped us off at the point where he picked us up. It was there that he held his hand out for a tip. In view of the pain, I was suffering in my back, neck and knuckles, I decided to give him 100 baht less my chiropractors fee which amounted to 90 Baht and we bade him farewell.

His face was one of sheer puzzlement-but hey, we were taller than him and there were two of us (ready for a fight) so he left in a cloud of smoke.

Chapter 20
Worthy Mention 1 Mysterious Roadkill and Shania Twain

When we flew to Christchurch in New Zealand during the latter part of December 2004, the plan was that we would first tour the South Island, as far as Queenstown, before taking another flight up to Auckland in the North Island and tour there in the same fashion. After we had landed in Christchurch, we stayed for one night at The George Hotel before picking a rental car for our onward journey (and the rest of our visit) in which we used motels for our overnight stays. I also need to mention at this point that although it was December and in the middle of their summertime the weather was in fact cold and wet for the majority of the time spent there. We were to say the least, p***ed off.

We did however deviate from our original plan and headed north to a place called Kaikoura where we went whale spotting from a helicopter as the sea was too rough for the normal boat trip. This was recommended in all the guide books we had read so it seemed appropriate that we at least give it a go. Following a successful sighting of some sperm whales, we found ourselves a suitable motel for an overnight stay.

The following day, we set off south once again for the long drive across the Southern Alps. This route would take us through a place called Arthurs Pass in the national park of the same name. This route was quite precipitous in parts and we had to endure some torrential rain (en-route) that made our journey even more dangerous especially as visibility was down to a few metres. At one point, we nearly fell to our deaths, as the car slid towards the roadside, that would have resulted in a sheer drop down the mountain. Thankfully, that danger was avoided and we eventually reached the west coast where the weather picked up along with our spirits.

Now there are two discoveries that we made (whilst travelling on the South Island) that we would like to share with you at this juncture. The first of these is the high levels of roadkill that we saw as we journeyed on. This may not seem strange to you until I mention the fact that we hardly ever saw any other vehicles on the roads we were on. Subsequently, we came to the conclusion that there are two possible explanations for this high proportion of dead creatures. One is that everyone drove at night, or two, the creatures themselves committed suicide as soon as a vehicle was spotted by them. We never did find out.

The second discovery was (according to a local daily newspaper) that out of 234 road accidents that were recorded in 2003, the police were responsible for just over 200 of them as a result of high-speed car chases. The recommendations of the report were that all traffic officers were to be given additional training. Luckily for us, we did not see any police cars on our travels. Maybe this could be a third solution to the roadkill issue.

Anyway, I digress. After another overnight stop at a motel (on the west coast), we eventually reached our destination of Queenstown which is a beautiful town on the banks of Lake Wakatipu. We arrived there on Christmas Eve and settled into our pre-booked hotel which was on a bed and breakfast arrangement. Conscious of Christmas Day approaching, we tried to reserve a table for lunch at the hotel but were told that they were fully booked and suggested we try other establishments in the town. This we set about doing but were met with the same response, so we had to resign ourselves to buying a sandwich or whatever we could scavenge from the local back-packers café/shop which turned out to be the only establishment that was open on Christmas Day. If you're interested, I settled for a BLT whilst Mrs J had a prawn mayo. Well! It was Christmas after all.

We also learnt that Shania Twain had secretly flown into Queenstown for that Christmas and the word was that she wanted my autograph. (Well! What other reason could there be). This presented somewhat of a dilemma and we had to go around incognito for the rest of our time there. Luckily, our paths did not cross even though I knew she was a big fan. The local press however had discovered her whereabouts but not mine.

Undaunted we made the most of our time at this beautiful location but the weather still let us down. So much so that we decided to change our intended itinerary and give the North Island a miss due to a dodgy weather forecast for that particular zone. So, we drove to Queenstown Airport and arranged for our flight tickets to be amended for a small fee.

The new arrangement was that we would travel back to Sydney in Australia and spend some additional time there. After that, we resume our journey to Hong Kong as previously arranged.

Our flight out took place on 26 December 2004 and about halfway into the flight, we experienced severe air turbulence that lasted for several minutes. Although I cannot say for certain, this turbulence coincided exactly with the time of the Indonesian tsunami that we learnt about after landing in Sydney. Having said that, it could have been a near miss with a police helicopter in a high-speed chase. The jury is out.

Chapter 21
Worthy Mention 2
Have You Any Lipton?

When we arrived in Hong Kong, we were well pleased with the weather as it was much warmer. Our hotel (The Novotel) which had been pre-booked, was ideally situated with views over Victoria Harbour and was on a bed and breakfast arrangement. This suited us as we had more freedom to explore and sample some of the local cuisine whilst on our travels. Although no longer a British colony, it still had that colonial feel as The Republic of China had not yet exerted its full control over the island.

Believe it or not, Hong Kong has its very own China Town where you can buy all sorts of weird foods and general goods. When we paid a visit, we saw strange things being offered for sale. There were dried insects of just about every description, live fish and eels as well as raw meat that could have been rat for all we knew. The various vendors were crammed together in back alleys with what looked like a sewer system running between each set of business corridors and it was advisable to avoid stepping into whatever was coursing its way down these channels.

We also encountered small stalls selling such items as jewellery and knick-knacks although you could only buy items in multiple amounts of say a dozen at a time. This was wholesale with a difference. The other apparent thing about this area was the strange smell that emanated amongst these shopping areas and pedestrian footways. If I had to guess the smell, I would say "over boiled turnip meets sweaty-socks." Above all, it was oppressive and we were glad to leave the area and explore further afield.

There were however some other interesting places that we visited whilst there. One of these was a trip on the tram-type railway system that took tourists up to Victoria Peak, one of the highest points on Hong Kong Island. At the top, we took in some spectacular views and also availed ourselves of some typical Chinese food for our lunch. Afterwards, we decided to walk back down through what was a lovely park containing a huge aviary with tropical birds and a very old looking giant pelican. This walk turned out to be one of the highlights of our visit and I would recommend it to anyone.

During our continued visit around the area, we decided upon a ferry trip across the bay to Kowloon which was effectively mainland China. Our intention was to visit a famous floating restaurant that had been recommended to us for an excellent lunch. During the crossing, we could not help noticing that there was such a thing as marine bin-men. In reality, they were crews operating small boats that had the main task of picking up (with nets) any floating debris, that littered the water channels. We both thought that this was an excellent idea but wondered where they deposited their collections of muck. That's the trouble with China as you just never know.

Having sampled an excellent Chinese lunch feast, our appetites were wetted for an authentic breakfast before we departed this exotic location. After diligent enquiries (the following day), we were directed to a tower block not far from our hotel. We were told that we would be able to recognise the spot by the queue of Chinese people congregating around a lift door that took customers up to a tenth-floor restaurant. Sure enough, we spotted our target and accordingly joined the throng waiting for the next available lift. I have to say at this point that we were the only westerners looking to join the hungry crowd and were starting to feel a little out of our depth.

Undaunted we travelled in the crowded elevator up to a lobby that led to a set of double doors. Following the crowd, we entered into a large dining room full of local residents (having breakfast) who suddenly went quiet and looked at us as if we were creatures from another planet. After a while, the novelty wore off and their conversations resumed as normal. An elderly lady (who we assumed was some sort of 'Maitre D') came over to greet us and showed us to a table. She handed us a menu each before departing to deal with other customers. To our horror, the menu was written purely in Chinese but luckily, had a few pictures of what was on offer.

Now I don't know about you, but when I am placed in a situation where no one speaks English, I naturally resort to my time-honoured slow drawl method that is accompanied by hand gestures to get my message across. This generally makes Mrs J want to pull her hair out. However, when the time came for us to place our order, we had to point at the various pictures and hold up one or two fingers to denote the quantity we wanted. This seemed to work until the waitress (pointing at the cups on the table) enquired as to what we wished to drink with our breakfast. I knew we wanted some tea but was at a loss for what to ask for. Making a 'T' shape with my two forefingers was no good so I had to resort to a description of the tea bags we had in our room at the hotel.

The drawn out reply I gave went something like this. "Have-you-any-Lip-Ton?"

I could sense Mrs J cringing but the waitress seemed to grasp what I was trying to convey and (as if she thought I naturally talked like this) she replied, "Noh! On-Ly-Green-T." At last, someone who understood me. So! I replied, "O-Kay-Two-Green-T," and held up the customary two fingers to confirm our order. Anyway, to cut a long story short, we ended up having a wonderful authentic Chinese breakfast, the memory of which lives long in my memory. *He who dares, wins*, as they say.

Chapter 22
Final Conclusions

And so, dear reader, I have reached a conclusion in my literary offering but feel I need to finish with some general advice as to taking holidays with a wife. First of all, you must have a wife. If not, see if you can borrow one from someone who you can trust. Secondly, pick your holiday friends with care, as a mistake here can lead to some unexpected events. Finally, choose a destination that suits you both so that you can let your hair down and relax. It goes without saying that relaxing with your hair down leads to fun, and fun usually leads to laughter. If you do however want to be awkward and do it with your hair up, then your guess is as good as mine.

We feel extremely lucky to have had both the time and the money to make the journeys that we have undertaken and also to have had the very good fortune to meet so many great (like minded) people. The events that followed these encounters will live in our memories for a very long time.

There is an old saying that goes *You can take a wife to water, but you can't make her drink it*. That is also the case with Mrs J who prefers alcohol every time, as she insists water is for washing and swimming in. You can't really argue with that and I learnt long ago that its best to keep your thoughts to yourself. Anyway, I am digressing again (which is probably due to my age) and I need to get back to the nitty-gritty of holidays.

Writing about 'taking a holiday with a wife' has been a total pleasure and stirred up some amusing recollections from my earlier life. The stories outlined in this book have been made possible thanks to some brief note-taking, at the relevant times. These have, in turn, permitted me to relate them as true-life adventures in a more narrative format.

I have to admit that I am no travel writer (or ever intend to be) so I hope you will forgive me for any omissions or inaccuracies in respect of the places mentioned on our travels. Instead, I have selfishly concentrated more on the humorous events that befell us, mainly because this book was never meant to be a travel guide per se.

As mentioned earlier, there were many more journeys undertaken during our marital years but due to lack of note-taking, my memory has failed me. Maybe, just maybe (now that the pandemic is almost over), new travels can commence and, with the assistance of modern technology, new memories can be collected and used as future stories to be told.

The End